Handbook of Biochemistry Practical Of Protein

By

Dr. Ekta Prakash
D.Phil (Biochemistry)
Allahabad University

SECTION I

S.No	CHAPTER	Page
1	Protein	6
2	Primary Stucture	6
3	Secondary Stucture	6
4	The α –helix:	7
5	The β- pleated Sheet:	7
6	Quaternary Structure:	7
7	Classification OF Protein	11
8	Simple Proteins	11
9	Soluble in Distilled Water	11
10	Albumens:	11
11	(b) Pseudoglobulins	12
12	(c) Protamines:	12
13	(d) Histones	12
14	2. Insoluble in Distilled Water	13
15	(a) Glutelins:	13
16	(b) Protamines:	13
17	(c) Globulins	13
18	B. Simple Fibrous Proteins:	13
19	1. Keratins	13
20	2. Collagen	13
21	3. Fibroin	13
22	II Conjugated Protein:	14
23	1. Chromoprotein:	14
24	1. Chromoprotein:	14
25	2. Glycoproteins:	14
26	3. Nucleoproteins:	14
27	4. Lipoproteins:	14
28	5. Phosphoproteins:	14
29	Properties of Proteins	14
30	1. Colloids:	15
31	2. Solubility:	15
32	B. Chemical Properties:	15
33	1. Chromoprotein:	15
34	2. Glycoproteins:	15
35	3. Nucleoproteins 4.Lipoproteins:	15 15
36	5. Phosphoproteins:	15
37	Properties of Proteins:	15
38	Amphoteric Naturen:	15
39	Coagulation	15

40	Optical property:	15
41	Hydrolysis	15

SECTION -II

1	EXPERIMENT No. 1 Biuret Test	16
2	EXPERIMENT No.2 Xanthoproteic Acid Test	18
3	EXPERIMENT No.3 Ninhydrin test	21
4	EXPERIMENT No. 4 Millon's Test	24
5	EXPERIMENT No. 5 Hopkins cole test	25
6	EXPERIMENT No. 6 Sakaguchi Test.	27
7	EXPERIMENT No. 7 Histidine Test	29
8	EXPERIMENT No. 8 Estimation of protein by Lowry method et al.	32
9	EXPERIMENT No. 9. Estimation of Protein by Biuret Method:	40
10	EXPERIMENT No10. The isolation of casein from milk	43
11	EXPERIMENT No 11.Estimation of Protein by Bradford protein assay Introduction.	44
12	EXPERIMENT No.12 Estimation of Protein By KJELDAHL method	46
13	EXPERIMENT No. 13 Estimation of Protein By Enhanced Duma method	49

SECTION III

1	Beer-Lambert Law	55
2	Principle of the Colorimeter	56
3	Reagents Required:	60
4	List of Material Required	63
5	Precaution in Lab	70

S.No	Figure	Page
1	Classification of Protein On the Basis OF Structure	8
2	Secondary structure with example	8

No.	Title	Page
3	Quaternary Structure	9
4	Primary structure	9
5	Secondary Structure	10
6	Quarnery Structure	10
7	Classification of protein on the basis of Solubility	11
8	Water soluble protein	12
9	Formation of BIURET -COMPLEX	16
10	Biuret Test	17
11	Structure of Trytophan and Tyrosine	18
12	Formation of orange- yellow complex	19
13	Xanthoproteic acid	19
14	Amino acids react with ninhydrin Rhuemann's purple	21
15	Amino acid Ninhydrin	22
16	Structure of Tyrosine	23
17	Millons Test	24
18	Structure of TRYPTOPHAN	25
19	Hopkin Cole Test	26
20	Structure arginine	27
21	Sakaguchi Test	28
22	Structure of Histidine	29
23	Histidine Test	30
24	Reaction showing formation of Blue colored complex by Folin-Ciocalteau reagent in Lowry'method	30
25	Estimation of Phenolic compound in Biological sample by Lowry Method	32
26	Estimation of Protein Taking BSA as Standard by Lowery et al Method	37
27	Estimation OF Protein OF Latex by Lower et. al Method	38
28	Estimation OF Protein by Lower et. al Method	39
29	Estimation of protein by Biuret Method at different conc	41
30	Estimation of protein by Biuret Method	
31	Estimation of Protein by Bradford protein assay	44
32	Bradford Reagent	
33	Graph depicting Beer-Lambert Law	54
34	Visible Light Spectrum from wavelength 400-700nm	55
35	Different color of light absorb different wavelength of light	56
36	Beaker	63
37	Test Tube	63
38	Flat Bottom Conical Flask	64
39	Test Tube	64
40	Pipette	65
41	Test Holder	65
42	Colorimeter	66
43	Vortex	66
44	Waterbath	67
45	Pipette	67

| 46 | Analytical Balance | 68 |
| 47 | Oven | 68 |

S.No	Table	Page
1	Summary Of Protein Test	31
2	Optical Density standard	34
3	Optical Density of Biological Sample	35
4	O.D at Different Concentration	41

Chapter I

Proteins

Protein molecules are very large and elaborate polypeptide chains being formed by the linkage of several thousand molecules of amino acids. The chains remain folded and exhibit different structures. The chains remain folded and exhibit additional linkages between their amino acids. Therefore these exhibit different structure. In the structure protein molecules exhibit different levels of organization. These are primary, secondary, tertiary and quaternary. The concept of these levels of organization can be explained as under.

1. The peptide linkages and individual amino acids along the protein chains present the primary structure.

2. The configuration of the chain i.e. the coils in the chains represent its secondary structure.

3. Three dimensional configurations showing the bending entwining of these coils with each other in the chain represents the tertiary structure.

4. The number and nature of such entwining units in the entire protein molecule represents the quaternary structure

1. **The Primary Structure:**

The number, nature and sequence of amino acid molecules in a polypeptide chain is called the primary structure of protein. Since each protein molecules is formed of several thousand amino acid molecules the determination of primary structure of protein molecule is an extremely difficult task. **Sanger** (1955), determined the sequence of amino acid molecules of insulin for the firt time.

2. **Secondary Structure:**

The secondary structure of a protein is determined by hydrogen bonding between the carbonyl and amide hydrogen atoms of the component amino acids of the peptide chain itself. These bonds can occur either between different polypeptide chains of the protein or within the molecules of one polypeptide chain.

Three basic types of helix are used to define secondary structure. These helices are: α –helix, and β- helix

(i) **The α –helix:** In α –helix the peptide chain assumes the form of a spiral staircase with 3 ½ molecules of amino acids per turn. The coils of the helix are held together by hydrogen bonds lying parallel to the main axis of the fibre. Such a structure is flexible and elastic but it gives stability to the peptide chain.

When chain is extended the hydrogen bonds rupture, but these are reformed when the chain is allowed to relax. The α –helix are found in proteins such as muscle protein myosin and α –keratin in hair and nails.

(ii) **The β- pleated Sheet:**

In this type of structure, hydrogen bonding occurs between the peptide chains and the bonds are laid down at right angles to the main chains. Fibre containing the type of structure is not elastic because the polypeptide chains are already in extended form, but these are very strong and flexible.

Silk fibres are best example of β configuration; β keratin found in feathers and claws is another example of this type of protein.

4. Quaternary Structure:

The quaternary structure involves the noncovalent association of one or more peptide chains. For example a molecule of hemoglobin has two chains (α-chain and β-chain).

The quaternary structure can be two types.

(i) In which the protein molecules is formed of several dissimilar peptide chains but only one active site.

(ii) In which the protein molecule is formed of similar or identical subunits.

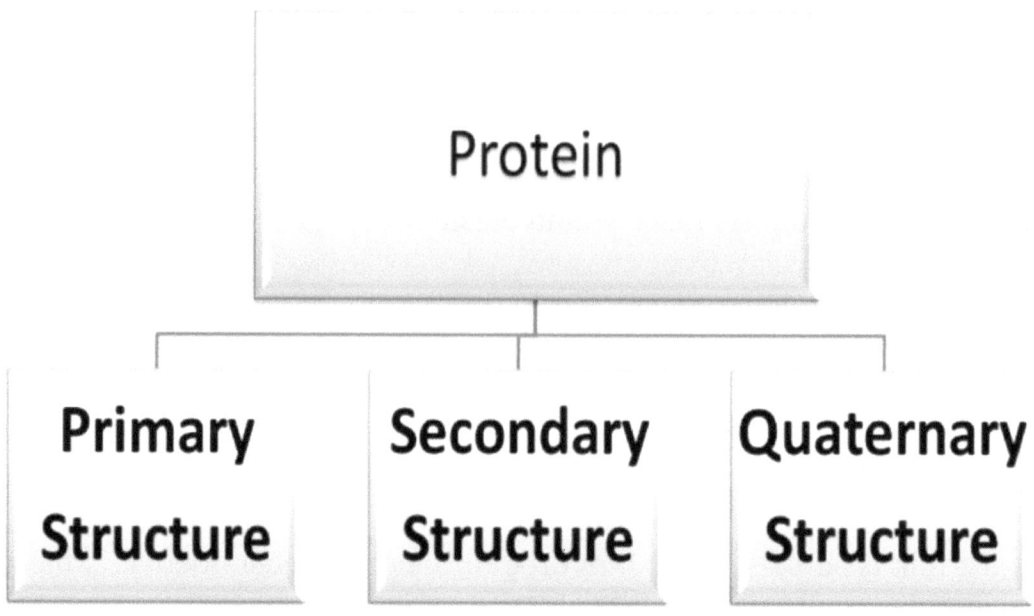

Fig 1. Classification of Protein On the Basis OF Structure

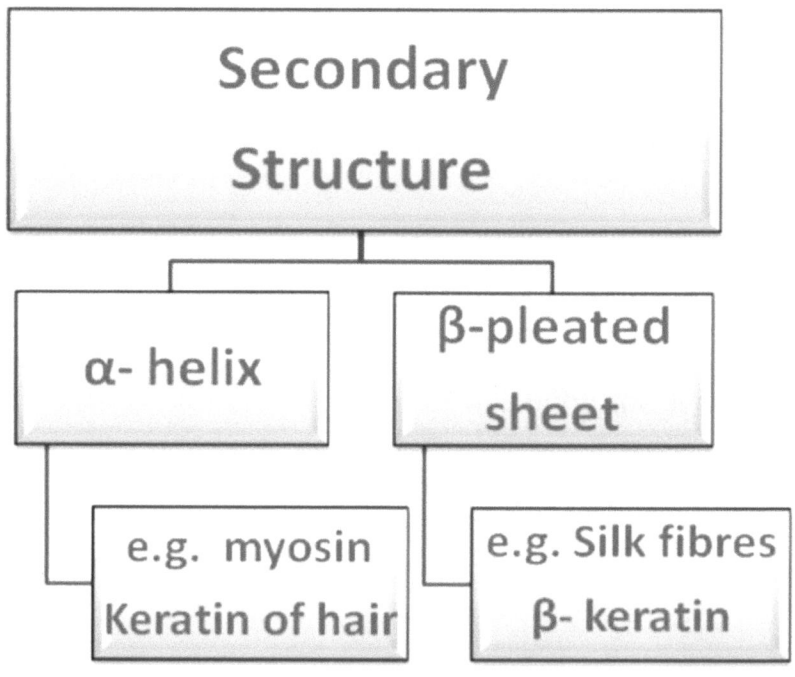

Fig 2. Secondary Structure With Example

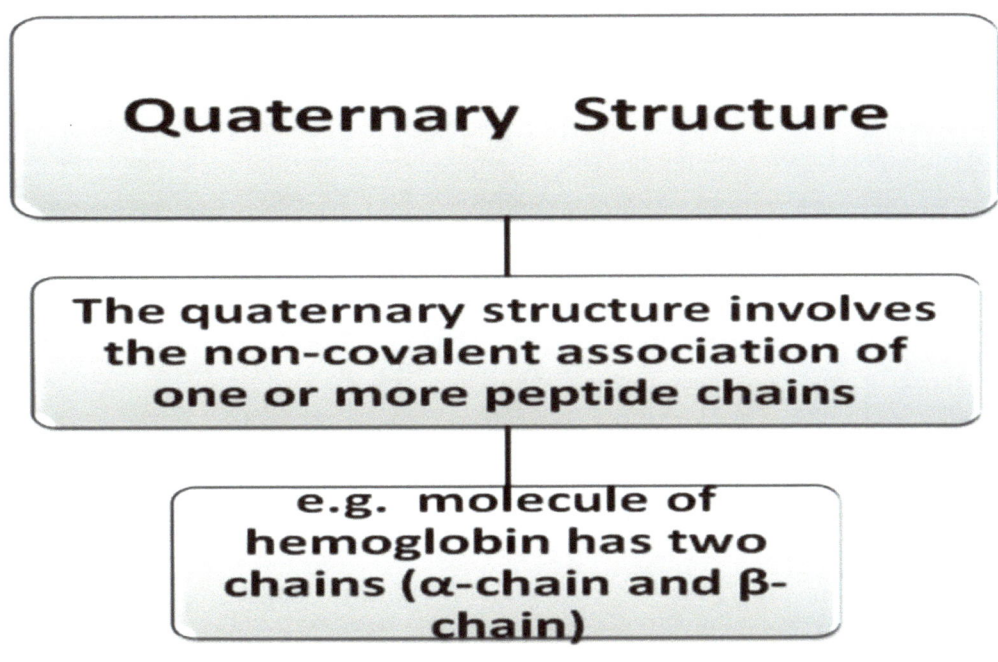

Fig 3. Quaternary Structure

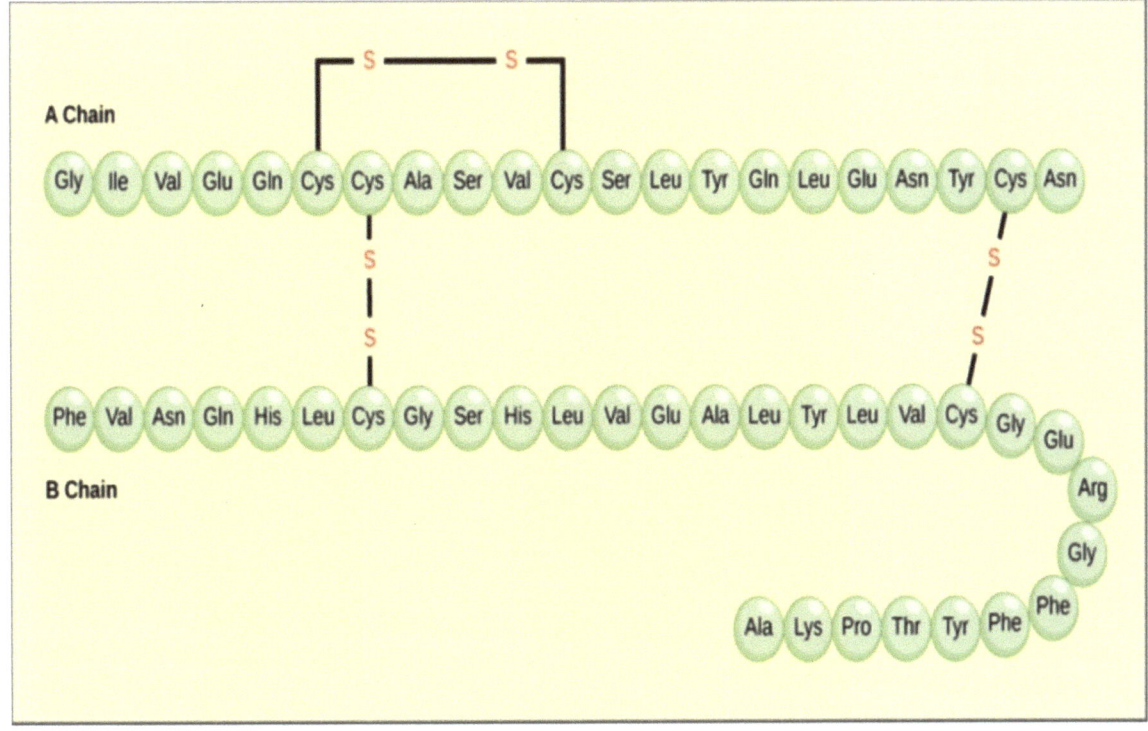

Fig4. Primary structure: The A chain of insulin is 21 amino acids long and the B chain is 30 amino acids long, and each sequence is unique to the insulin protein.

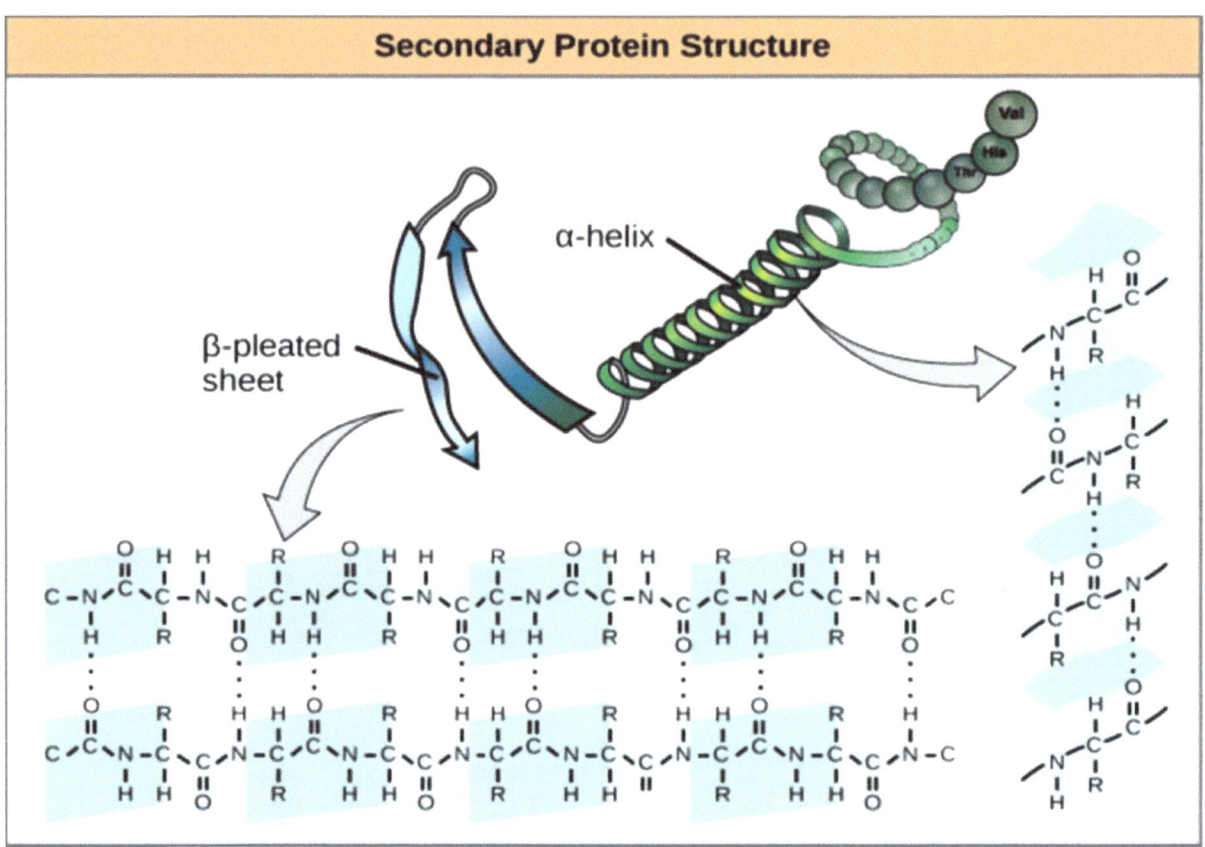

Fig 5. Secondary structure: The α-helix and β-pleated sheet form because of hydrogen bonding between carbonyl and amino groups in the peptide backbone. Certain amino acids have a propensity to form an α-helix, while others have a propensity to form a β-pleated sheet.

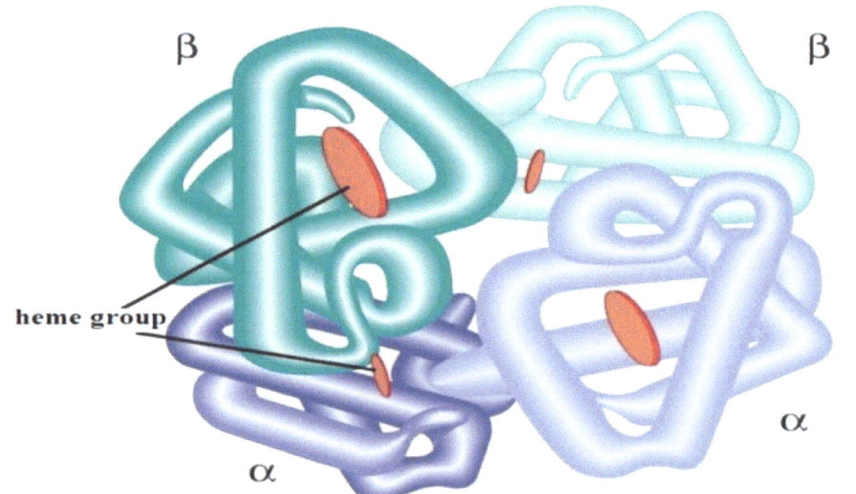

Quaternary Protein Sructue: Three-dimensional assembly of subunits

Fig 6. Quaternary Structure

Classification OF Protein:

Proteins have been classified on the basis of structure, solubility and coagulability. According to this classification proteins are separated into three categories.

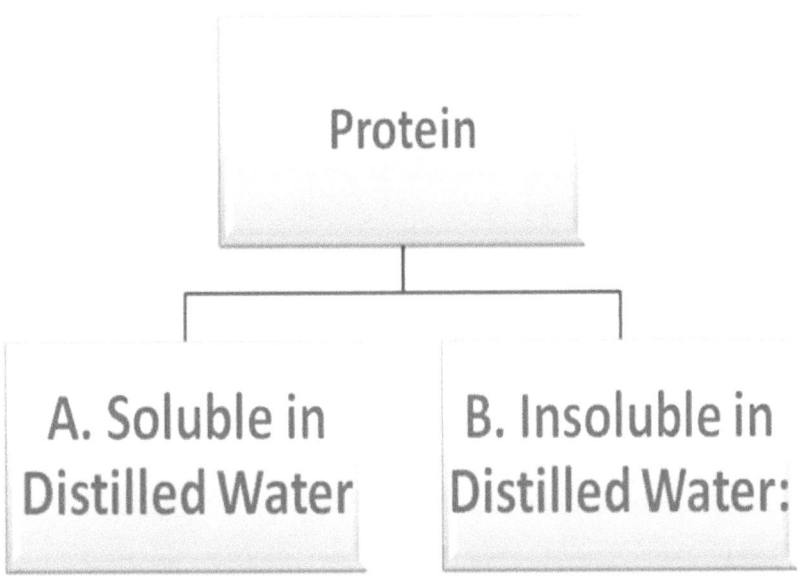

Fig 7. Classification of protein on the basis of Solubility

1. Simple Proteins

Simple protein on hydrolysis yield only amino acids

A. Soluble in Distilled Water

(a) Albumens:

- These are precipitated from water by dilute acids and alkalies.
- These can be precipitated by even saturating the solution either with a neutral salts such as NA_2SO_4 in slightly acidic solution or acid salt such as $(NH_4)_2SO_4$.
- In presence of stronger acids and alkalies these are converter into soluble metaproteins.
- On heating albumens get coagulated.

- Albumens are widely distributed in nature. Examples are egg white ova albumen, blood serum albumen

(b) Pseudoglobulins:

These can be precipitated from water solution with an acid salt $(NH)_2SO_4$. These are rare in nature. An example is the pseudoglobulin of milk whey.

(c) Protamines:

- These are basic proteins highly soluble in water and in dilute acids and dilute ammonium hydroxide solution.
- These form crystalline salts with mineral acids and insoluble salts with more acidic proteins. These are not coagulated by heating.
- These are simplest of all the naturally occurring proteins and have low molecular weight.
- Protamines are isolated from mature sperms

(d) Histones:

- These are basic proteins of high molecular weight.
- These are soluble in water and dilute mineral acids but not in ammonia hydroxide.
- This occur as part of nucleoproteins.

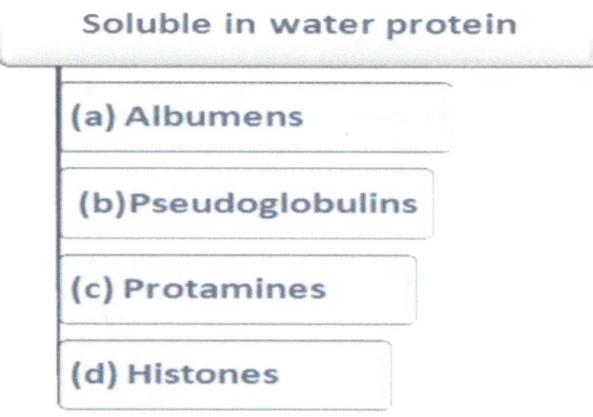

Fig 8. Water soluble protein

2. Insoluble in Distilled Water:

(a) Glutelins:

- These are insoluble in distilled water and neutral salt solution but are soluble in dilute alkalis and acids.
- These are found exclusively in seeds of cereals grains.

(b) Protamines:

- These are insoluble in water but are soluble in dilute alkalies and 60%-80 % alcohol.
- These are found in plants only. Gliadin, zein(maize).

(c) Globulins:

- These proteins are insoluble in water but readily soluble in dilute neutral salt solution such as NaCl.
- These coagulate on heating and heat coagulation is expected by the addition dilute acids. Egg globulin or vitellin of egg yolk, fibrinogen of blood plasma myogen and myosinogen of the muscle are animal globulins.
- Vegetable globulins include legumins.

B. Simple Fibrous Proteins:

- On account of fibre like molecular structure the fibrous proteins are insoluble in cold water and any other cold reagent.
- These are also known as sclerproteins. These are found exclusively in animals.

Examples of simple fibrous proteins are:

1. Keratins- These are found in the outer layer of skin and in hair, feather horns, hoof and nails.

2. Collagen: Collagen proteins are found is white fibrous connective tissue which constitutes tendons and fascia. These found ground substance of elastic cartilage. It is insoluble and hard to digest.

3. Fibroin: This protein is present in silk

II Conjugated Protein :

These proteins are composed of simple protein united with some nonprotein substance. This nonprotein group is known as "prosthetic group". For example in haemoglobin proteins globin is combined with an iron containing compound, heme. Depending upon the nature of prosthetic group the conjugated can be classified.

Proteins have been divided into several classes:

1. Chromoprotein:

- The simple protein is combined with a pigment for example for haemoglobin, cytochromes and flavoproteins.

2. Glycoproteins:

- In glycoproteins the simple proteins are combined with carbohydrate. For example mucin of saliva.

3. Nucleoproteins:

- In these proteins the proteins molecules are combined with nucleic acids.
- These proteins are protamines or histones.
- The chromatin material of the nuclei of cells is composed of nucleoproteins.

4. Lipoproteins:

Liproteins are formed by the combination of simple proteins with lipids. These proteins are located in the brain, plasma, egg and milk etc.

5. Phosphoproteins:

- Phosphoprotein are the proteins in combination with phosphoric acid or with ortho or pyrophosphate.
- These are soluble in dilute alkalies and are precipitated by the addition of acids.
- Ovovitelline of egg and caseinogens and casein of milk are example of phosphoproteins.

Properties of Proteins:

1. **Colloids**:
Proteins exists in colloidal state. Their molecules are of large size; therefore, these are unable to diffuse through plasma membrane.

2. **Solubility**:
Because proteins are colloids of large sized molecules, these form turbid solution in water. These are insoluble in alcohol. These are precipitated by acids in certain concentrations.

B. Chemical Properties:

Amphoteric Nature:

Proteins are amphoteric in nature. These behave as acids to alkaline solution and alkaline to acidic solutions and form salts with them.

Coagulation:

On heating proteins are coagulated but temperature of heat for coagulation differs for different for different proteins.

Optical property:

Amino acids are optically active except glycine. These are mostly laevorotatory. Some of the conjugated proteins haemoglobins and nucleoproteins are dextro- rotator.

Hydrolysis:

When proteins are boiled with dilute mineral acids in a reflux condenser, the protein molecules gradually break up into simple ones until these are reduced to amino acids hydrolysis occurs in the following steps.

Proteins ⟹ Proteoses ⟹ Peptones ⟹ Amino Acids

Experiment No. 1

Biuret Test

(i) Objective : The Biuret Test for the Peptide Bonds

(ii) Principle:

Alkaline copper sulphate reacts with compounds containing two or more peptide bonds to give a violet colored complex. The depth of the color obtained is a measure of the number of peptide bonds present in protein. The name of test comes from the compound biuret which gives a typical positive reaction. The reaction is not absolutely specific for peptide bonds, since any compound containing two carbonyl groups linked through nitrogen or carbon atom will give a positive result.

(iii) Reaction:

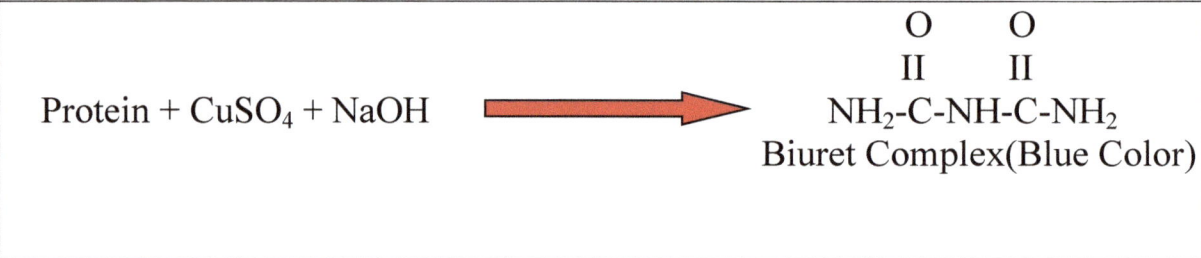

$$\text{Protein} + \text{CuSO}_4 + \text{NaOH} \longrightarrow \underset{\text{Biuret Complex (Blue Color)}}{NH_2-\overset{\overset{O}{\|}}{C}-NH-\overset{\overset{O}{\|}}{C}-NH_2}$$

Fig 9: Formation of Biuret complex when protein reacts with copper sulphate and sodium hydroxide

(iv) Reagent:

1. Copper sulphate (1%): 250 ml

2. Sodium Hydroxide (40%): 2L

3. Proteins (5g/l albumin, casein, gelatin; casein dissolved in a little dilute NaOH and other protein in saline): 500 ml

(v) Method:

Add five drops of copper sulphate solution to 2 ml of the test solution followed by 2 ml of NaOH; mix thoroughly and note the colours produced.

> Test solution + Copper sulphate + Sodium hydroxide ⟶ Blue Color
> Biuret Complex

Fig 10: Biuret Test for the Peptide Bonds

(vi) Observation:

Violet color is produced

(vii) Result:

Compounds containing peptide bonds produces characteristic blue color. All amino acid give this reaction.

Experiment No. 2

Xanthoproteic Acid Test

(i) Objective: To Test the Presence of Aromatic Amino Acid by Performing **Xanthoproteic Acid Test**.

(ii) Principle:

Aromatic amino acids, tyrosine and tryptophan, respond to this test. In the presence of concentrated nitric acid, the aromatic phenyl ring is nitrated to give yellow colored nitro-derivatives. At alkaline pH, the color changes to orange due to the ionization of the phenolic group.

(iii) Reactions:

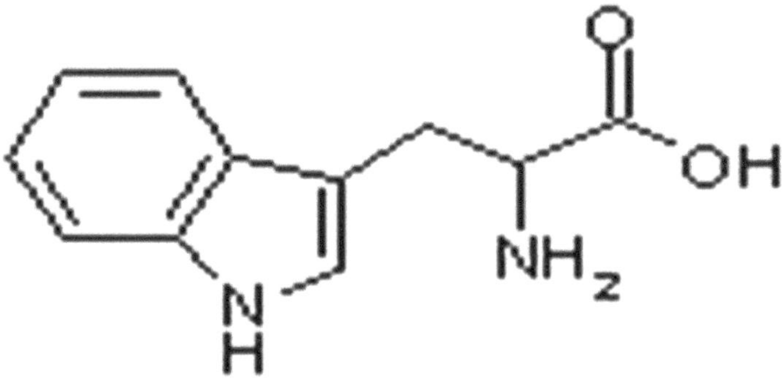

Tryptophan

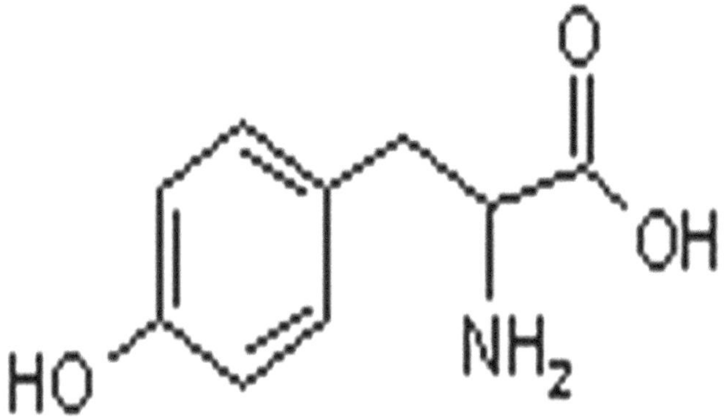

Tyrosine

Fig 11. Structure of Trytophan and Tyrosine

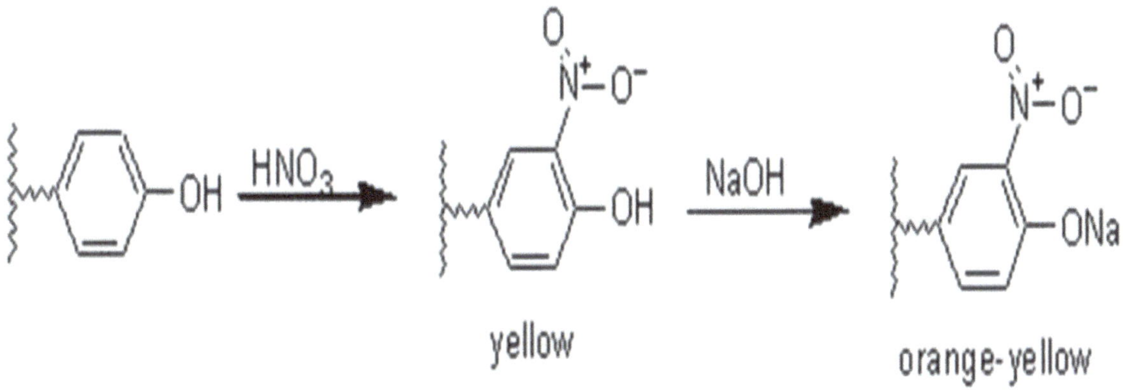

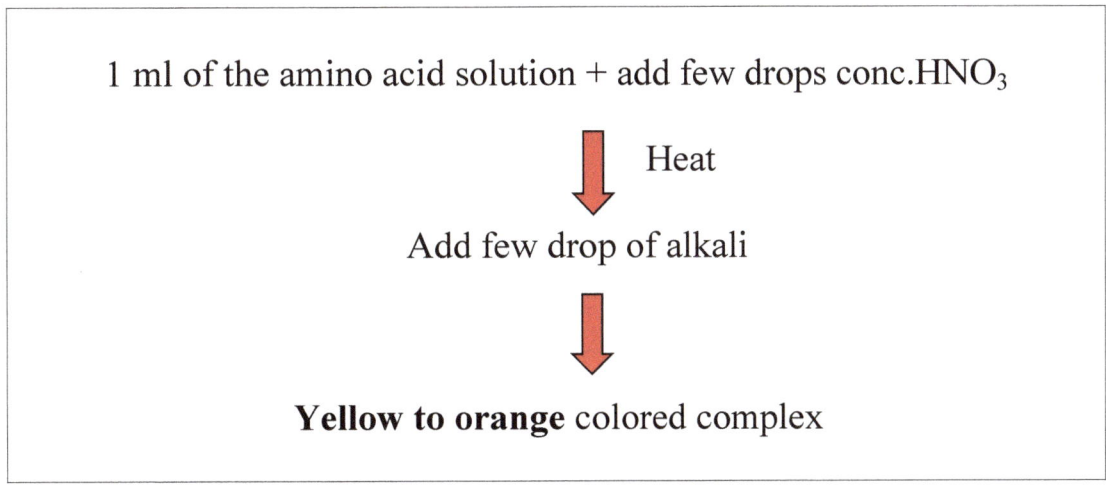

Fig12. Formation of orange- yellow complex

(iv) Reagents:

- Conc. Nitric acid
- Alkali, NaOH 40%

(v) Xanthoproteic acid Test :

To 1ml of the amino acid solution taken in a test tube, add few drops of nitric acid and vortex the contents. Boil the contents over a Bunsen flame, using a test tube holder, for few minutes. Cool the test tube under running tap water and add few drops of alkali.

> 1 ml of the amino acid solution + add few drops conc.HNO_3
>
> ↓ Heat
>
> Add few drop of alkali
>
> ↓
>
> **Yellow to orange** colored complex

Fig 13: Xanthoproteic acid Test

(vi) Observation:

Formation of orange yellow color complex solution seen

(vii) Result:

Aromatic amino acid is present in solution. The yellow color develops on boiling with conc. HNO_3 due to the presence of benzene ring. This reaction is due to the nitration of the phenyl ring of tyrosine, tryphtophan and phenylalanine.

Experiment No. 3

Ninhydrin test

(i) Objective: To test the presence of amino acid by **Ninhydrin test**

(ii) Principle:

In the pH range of 4-8, all α- amino acids react with ninhydrin (triketohydrindene hydrate), a powerful oxidizing agent to give a purple colored product (diketohydrin) termed **Rhuemann's purple**. All primary amines and ammonia react similarly but without the liberation of carbon dioxide. The **imino** acids **proline** and **hydroxyproline** also react with ninhydrin, but they give a **yellow colored complex** instead of a purple one. Besides amino acids, other complex structures such as peptides, peptones and proteins also react positively when subjected to the ninhydrin reaction.

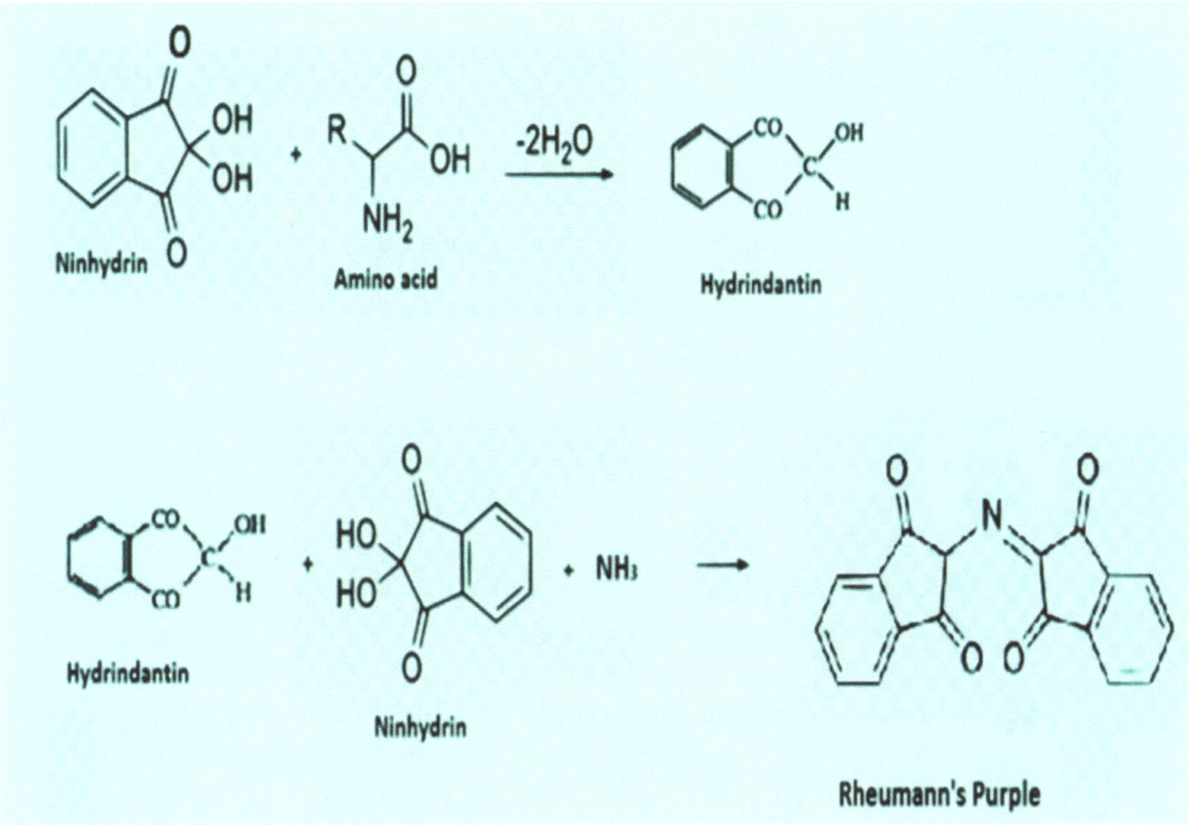

Fig 14: Reaction depicting formation Rheumann's Purple

(iii) Reagent:

Ninhydrin reagent: Ninhydrin (2 % W/V) in acetone

(iv) Ninhydrin Test:

To 1ml of amino acid solution taken in a test tube, add few drops of ninhydrin reagent and vortex the contents. Place the test tube in a boiling water bath for 5 minutes and cool to room temperature.

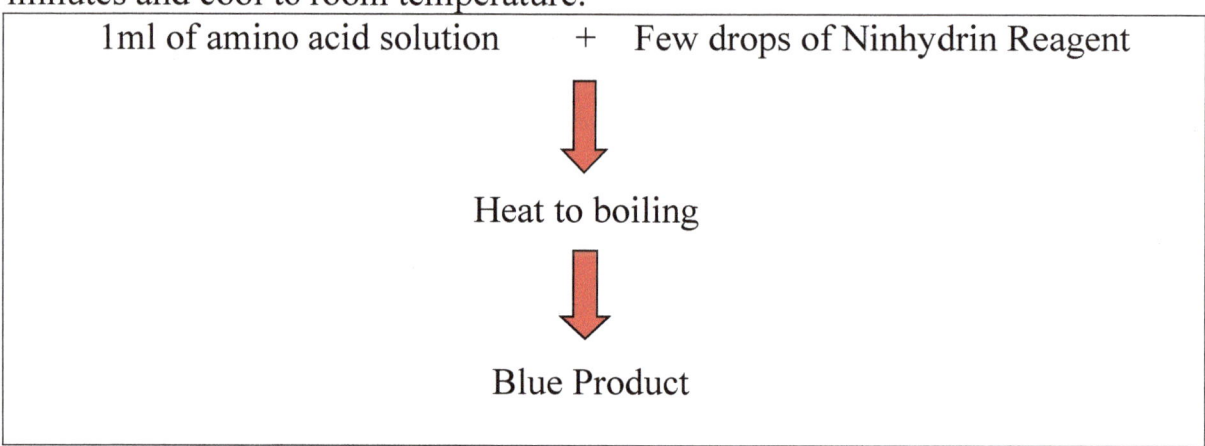

Fig15. Amino acids react with Ninhydrin triketohydrindene hydrate), to give a purple colored product (diketohydrin) termed Rhuemann's purple.

(v) Observation:

Purple colored complex solution is observed

(vi) Result:

Purple colored complex so formed proved that biological sample is amino acid

Note:

- When protein is boiled with ninhydrin, a blue color is produced due to the presence of the alpha amino acid.

- All proteins are positive except **proline**

- Ninhydrin (2,2-Dihydroxyindane-1,3-dione) is a chemical to detect ammonia or primary and secondary amines. When reacting these amines a deep blue color is evolved

Experiment No. 4

Millon's Test

(i) Objective: To determine the presence of Phenolic amino acids by Millon's Test.

(ii) Principle:

Phenolic amino acids such as Tyrosine and its derivatives respond to this test. Compounds with a hydoxybenzene radical react with Millon's reagent to from a red colored complex. Millon reagent to from a **red colored complex**. Millons's reagent to from a red colored complex. Millon's reagent is a solution of mercuric sulphate in sulphuric acid.

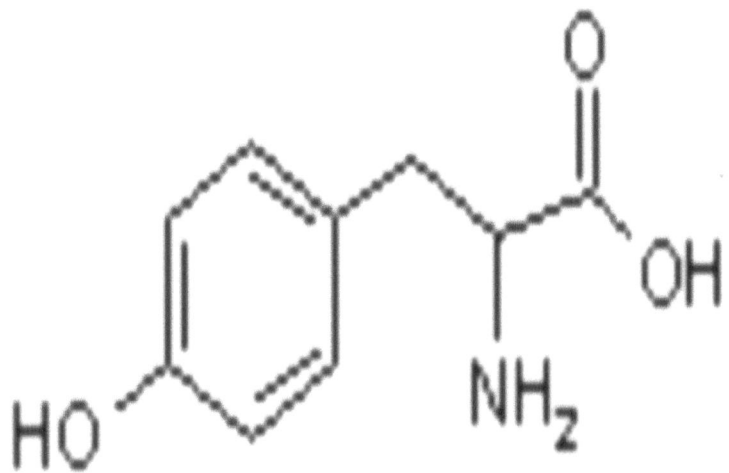

Tyrosine

Fig16. Structure of Tyrosine

(iii) Reagent:

1. Millon's reagent

2. Sodium nitrite solution (1%) [To be freshly prepared]

(iv) Preparation of Reagent:

7.5 gm of mercuric nitrate is dissolved in 50 ml of distilled
Water to make 15% mericuric nitrate solution 7.5 ml of conc. HNO_3 is diluted by 50 ml distilled H_2O to make 15% HNO_3 solution Now both the solutions are mixed together to prepare 100 ml.

(v) Millon's Test :

To 1ml of the amino acid solution in a test tube, add few drops of millon's reagent and vortex. Boil the contents over a Bunsen flame for 3 – 5 minutes. Cool the contents under running tap water and add few drops of sodium nitrite solution.

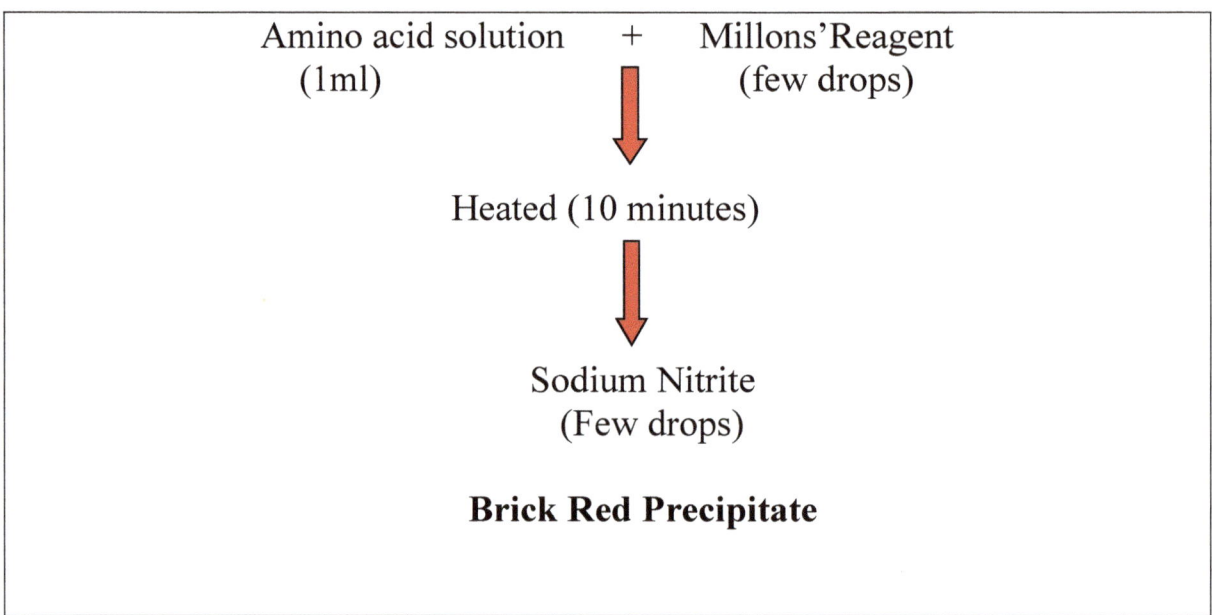

Fig 17. Millon's Test

(vi) Observation:

Red colored complex is formed

(vii) Result:

Red colored so formed proved that amino acid is present most probably tyrosine. The reaction is specific for tyrosine and takes place between mercuric and mercurous nitrate and tyrosine residue of protein. Tryptophan also responds to this reaction.

Experiment No.5

Hopkins cole test

(i) Objective: To detect tryptophan in solution by **Hopkins cole test**

(ii) Principle:

This test is specific test for detecting tryptophan. The indole moiety of tryptophan reacts with glyoxilic acid in the presence of concentrated sulphuric acids to give a purple colored product. A Glyoxilic acid is prepared from glacial acetic acid by being exposed to sunlight.

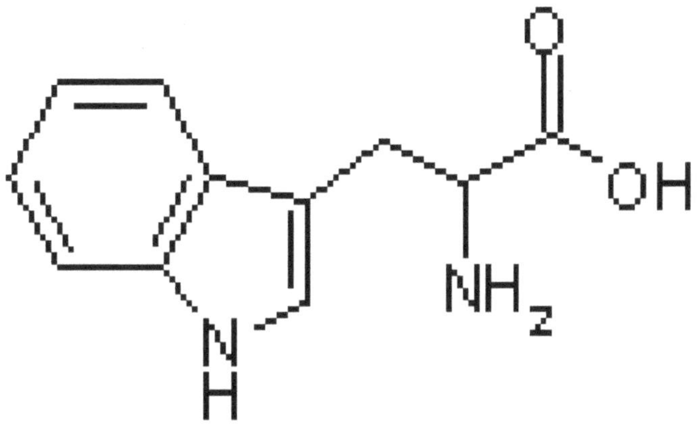

Tryptophan

Fig 18. Structure of Tryphtophan

(iii) Reagents:

Acetic acid – Glyoxilic acid reagent – Glacial acetic acid is exposed to sun light (for 5 – 6 hours) for the formation of small amounts of glyoxilic acid).conc. Sulphuric acid
Amino acids solution

(iv) Hopkins-Cole Test:

Mix 1 ml of the amino acid solution with 1 ml acetic acid glyoxilic acid reagent, in a test tube, vortex. Then carefully, add conc. Sulphuric acid along the side of the test tube, keeping the tube in an inclined position (do not shake the test tube, while adding acid).

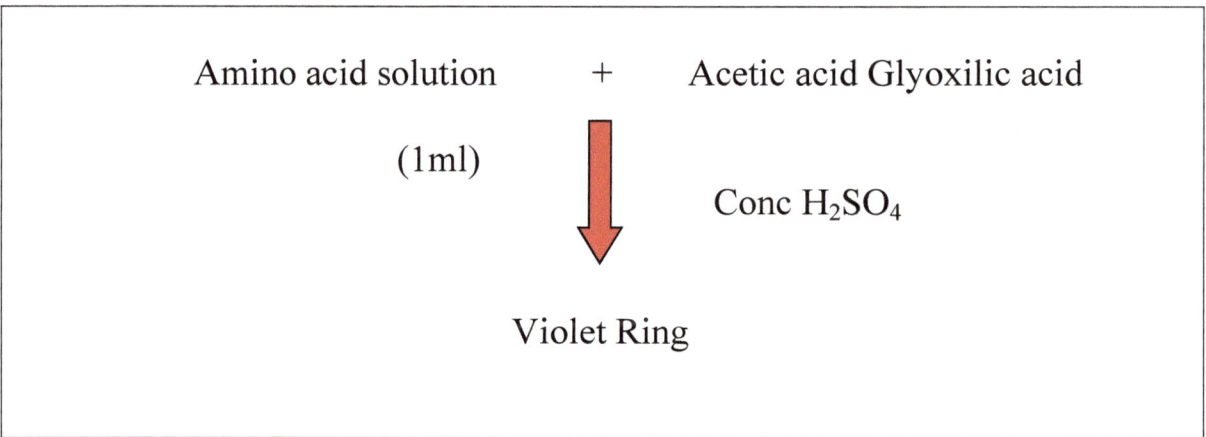

Fig 19. Hopkins-Cole Test

(v) Observation:

Purple colored solution complex is formed

(vi) Result:

Tryptophan is present in biological sample

Experiment No.6

Sakaguchi Test

(i) Objective: To determine arginine in biological solution by **Sakaguchi Test**.

(ii) Principle:

Under alkaline condition alpha naphthol (1-hydroxy naphthalene) reacts with a mono- substituted guanidine compound like arginine which upon treatment with hypobromite or hypochlorite, produces a characristic red color.

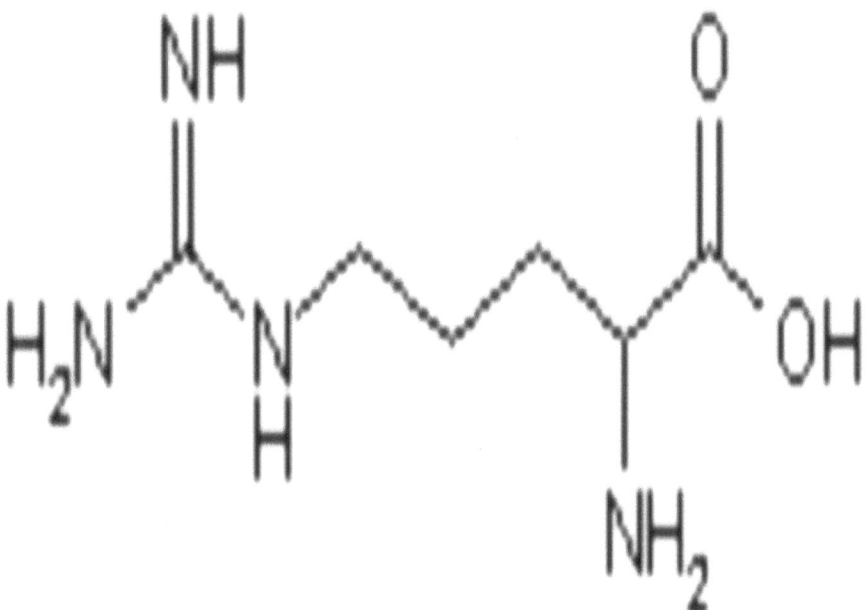

Fig 20. Structure of Arginine

(iii) Reagents:

1. Sodium hydroxide (10% W/V)

2. α Naphthol reagent (1% W/V in ethyl alcohol)

3. Hypobromite solution (To be freshly prepared) : Take 100 of 5%(W/V) sodium hydroxide solution in a glass reagent bottle and add 1ml of pre chilled liquid bromine, using a pro pipette. Shake the contents till bromine dissolves)

(iv) Sakaguchi Test:

To 1 ml of prechilled amino acid solution and few drops of NaOH is mixed and 2 drops of alpha naphthol is added. Mix thoroughly and add 4-5 drops of hypobromite reagent and observe.

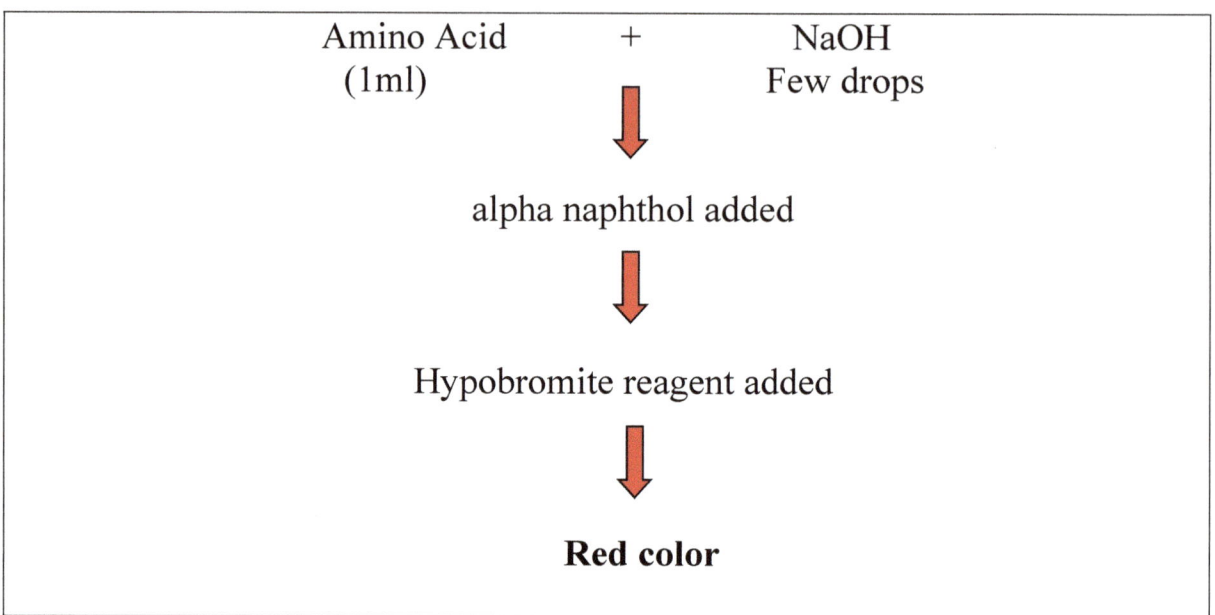

Fig 21.Sakaguchi Test

(v) Observation:

Red color solution is seen

(vi)Result:

Solution contain aginine. The red color so formed is suggest that biological sample is a mono- substituted guanidine compound like arginine is present.

Experiment No. 7

Histidine test

(i) Objective: To determine amino acid Histidine in biological solution by **Histidine Test**.

(ii) Principle:

The test was discovered by **knoop**. This reaction involves bromination of histine in acids solution, followed by neutralization of the acids with excess of ammonia. Heating of alkaline solution develops a blue or violet coloration.

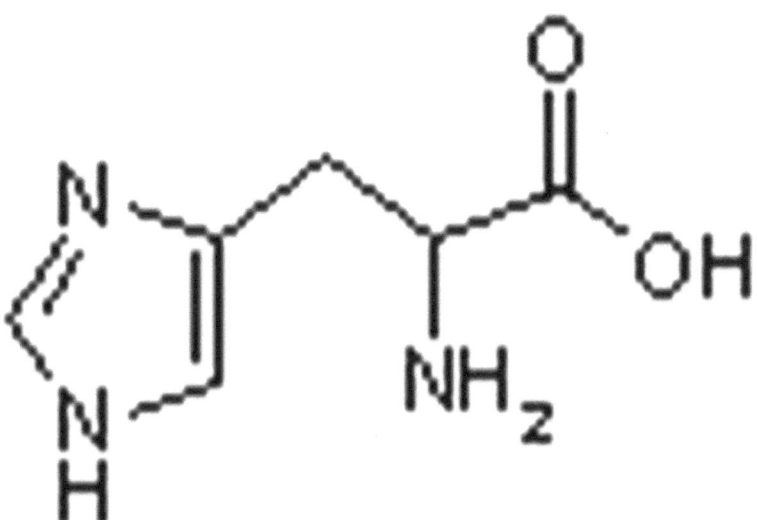

Fig22. Structure of Histidine

(iii) Reagents:

1. Bromine (5%) in acetic acid(33%) solution

2. Amonium carbonate (5%)

3. 33% acetic acid

(iv) Histidine Test:

To 1 ml of amino acid solution, add 5 % bromine in 33 % acetic acid until an yellow color was formed. After 10 minutes, add 2 ml of 5 % ammonium carbonate solution and placed in a boiling water bath for 10 minutes.

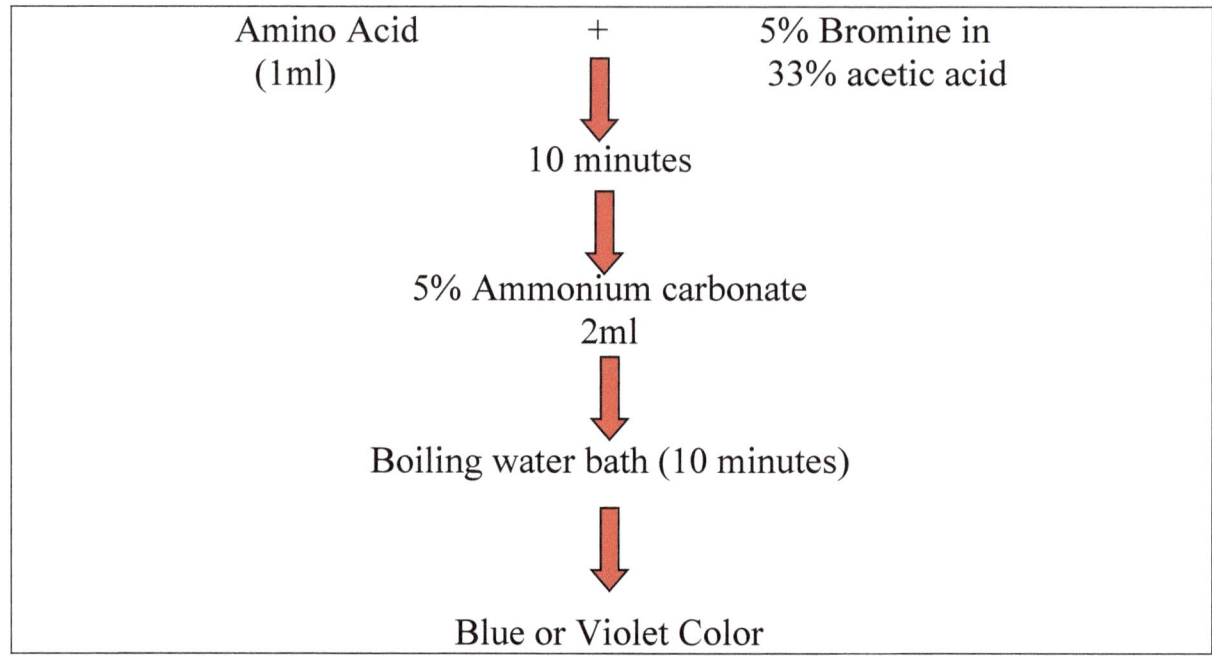

Fig 23. Histidine Test

(v) Observation:

Heating of alkaline solution develops a blue or violet coloration

(vi) Result:

Development of blue/violet color suggest that solution contains Histidine

S.No	Test	Composition of Reagent	Result (Color)	Group Responsible	Importance
1	Ninhydrin	Triketohydrin Hydrate	Blue or Purple	Free amino and COOH	Test for amino acid and peptides in dertmining amino acids
2	Biuret	NaOH + Cu_2SO_4	Violet	Peptide Linkage	Tripeptide upto protein
3	Millons'	Hg in HNO_3	Red	Hydroxy phenyl group	Tryptophan
4	Xanthproteic	Conc-HNO_3	Lemon Yellow	Benzene ring	Tyrosine, Phenylalanine, tryptophan
5	Hopkins'cole	Glyoxlic acid and conc. H_2SO_4	Violet Ring	Indole Group	Tryptophan
6	Liebermann	Conc.HCL, sucrose	Violet	Indole Group	Tryptophan, Histidine and tyrosine
7	Sakaguchi	10 % NaOH, α Napthol in alkaline hybobromide	Intense Red Color	Guanidine	Arginine

Table 1: Summary of Few Protein Test

Experiment No. 8

Objective: Estimation of protein by Lowry method *et al.* (1952) Modified

Principle:

Protein reacts with the Folin-Ciocalteau reagent to give a colored complex. The color so formed is due to the reaction of the alkaline copper with the protein as in the biuret test and the reduction of phosphomolybadate by tyrosine and tryptophan present in the protein. The intensity of colour depends on the amount of this aromatic amino acid present and will thus vary for different protein.

Lowary assay relies on two reactions:

1. Formation of a complex between Cu^{2+} and protein amide (peptide) bonds in an alkaline solution causing a reduction of copper to Cu^+ (Biuret reaction).

$$Cu^{2+} + \text{protein} \longrightarrow [Cu^{2+}\text{-protein complex}]$$

$$Cu^{2+} + (\text{polar amino acids, Trp, Tyr})\text{red} \longrightarrow Cu^+ + (\text{amino acids}) \text{ox}$$

2. Cu^+ and radical groups of tryptophan, tyrosine, and cysteine reduce a yellow phosphomolybdate-phosphotungstate complex (Folin-Ciocalteu reagent: $Na_2MoO_4 + Na_2WoO_4 + H_3PO_4$) to a deep blue color

$$Cu+ + (F\text{-}C)\text{ox} \longrightarrow Cu2+ + (F\text{-}C)\text{red}$$
$$(F\text{-}C) = \text{phospho-Mo-Tungstate acid}$$

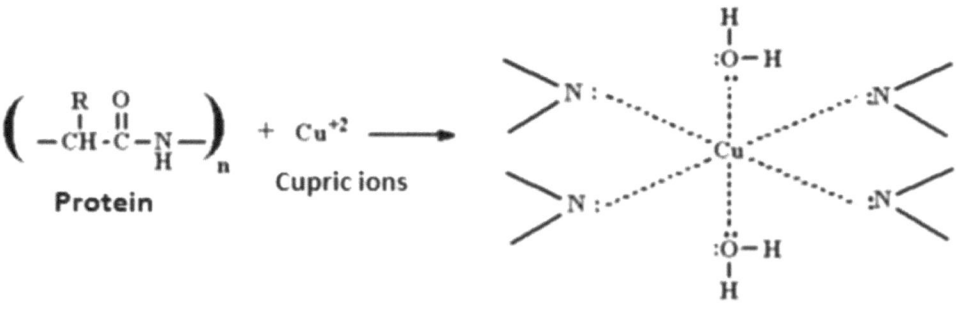

Complex formed

Fig 24. Reaction showing formation of Blue colored complex by Folin-Ciocalteau reagent in Lowry'method.

Preparation of reagent:

1. 0.1 N NaOH: Dissolve 4 g NaOH in glass distilled water and made the volume upto 1 liter with GD water.

2. Alkaline Copper Reagent:

(a) **Sodium Carbonate solution** (Na_2CO_3) (2%, w/v):   100 ml
Sodium carbonate 2g was dissolved in sodium hydroxide 0.1N and the volume made up to 100 ml

(b) **Copper sulphate solution (1%, w/v):**   100ml

Copper sulphate (1.0 g) was dissolved in glass distilled water and volume made to 100 ml.

(c) **Sodium Potassium Tartrate (Rochelle salt) (2%,w/v)** 100ml

Sodium potassium tartrate 2 g was dissolved in glass distilled water and the volume raised to 100 ml.

Alkaline copper sulphate solution was prepared by mixing 100ml of solution (a) with 1.0 ml each of (b) and (c).The solution was prepared fresh daily just before use.

3. BSA (Bovine serum albumin)

4. Stock solution (200 g /ml):

Dissolved 20 mg of BSA in 0.1N NaOH and made the volume upto 100 ml with 0.1N NaOH 0 .1 ml this contain thus 200 µg of BSA. We got 1 mg /ml solution of BSA.

5. Working solution:

Took 10 ml of this stock solution and diluted to 50 ml 0.1 N NaOH sodium hydroxide solution.

6. Folin's Reagent :

A mixture of sodium tunsgstate (50g) sodium molybadate (12.5 g), water (350 ml), phosphoric acid (50 ml, 85, w /v) and conc. Hydrochloric acid (50 ml) was refluxed in a one liter flask for 18 hour.

Then lithum sulphate (75 g), water (25 ml) And few drops of bromine water were added to the flask the mixture was boiled to Remove bromine vapor s, cooled to room temperature and diluted to 500 ml with glass distilled water. The reagent was titrated with sodium hydroxide (1.0N) to Phenolphthalein end point. On the basis of titration, the reagent was dilute with water to make it equinormal with sodium hydroxide. The reagent was then stored in a refrigerator. On the day of use took 36 ml of folin reagent and dilute it to 72 ml with glass distilled water. Folin reagent is light sensitive so it should be kept in brown bottle. And colour of reagent should check before use.

7. Experimental Procedure:

Suitable aliquots (0.2-1.0 ml) were pipetted into a series of tubes and the volume was raised to 1.0 ml with sodium hydroxide (0.1N). To each tube alkaline copper sulphate solution 95.0 ml was added, mixed well and allowed to stand at room temperature for 10 minutes. Folin's reagent (0.5 ml) was then added, the content of the tubes were mixed well immediately and after allowing to stand for 30 minutes, the optical density was measured at 660 nm. Reagent blank and standard solutions of protein (20-60 µg) were also run simultaneously for the calibration of curve.

A. Standard:

BSA (ml)	A620nm
00 ml	00
0.1	0.02
0.2	0.03
0.3	0.04
0.4	0.06
0.5	0.08
0.6	0.09
0.7	0.12

B. Biological Sample:

Biological sample (ml)	A620nm
00 ml	00
0.1	0.11
0.2	0.18
0.3	0.33
0.4	0.44
0.5	0.55
0.6	0.69
0.7	0.78

Table 1 & 2: Estimation Of Protein by Lowry Method

Estimation of Phenolic compound in Biological sample by Lowry Method   **Experimental**

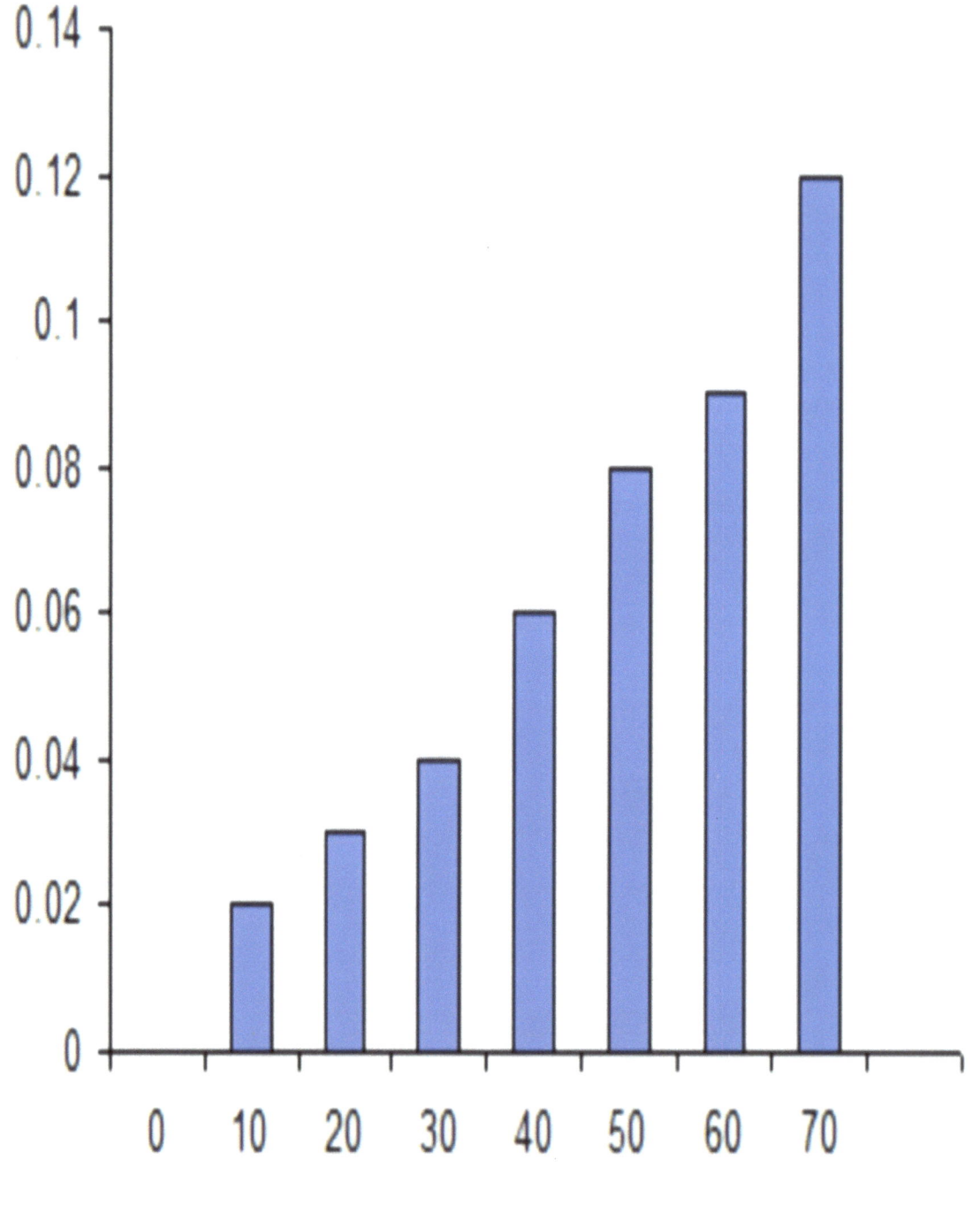

BSA conc. (ug)

Fig 25. Estimation of Phenolic compound in Biological sample by Lowry Method
A.. Standard
B.

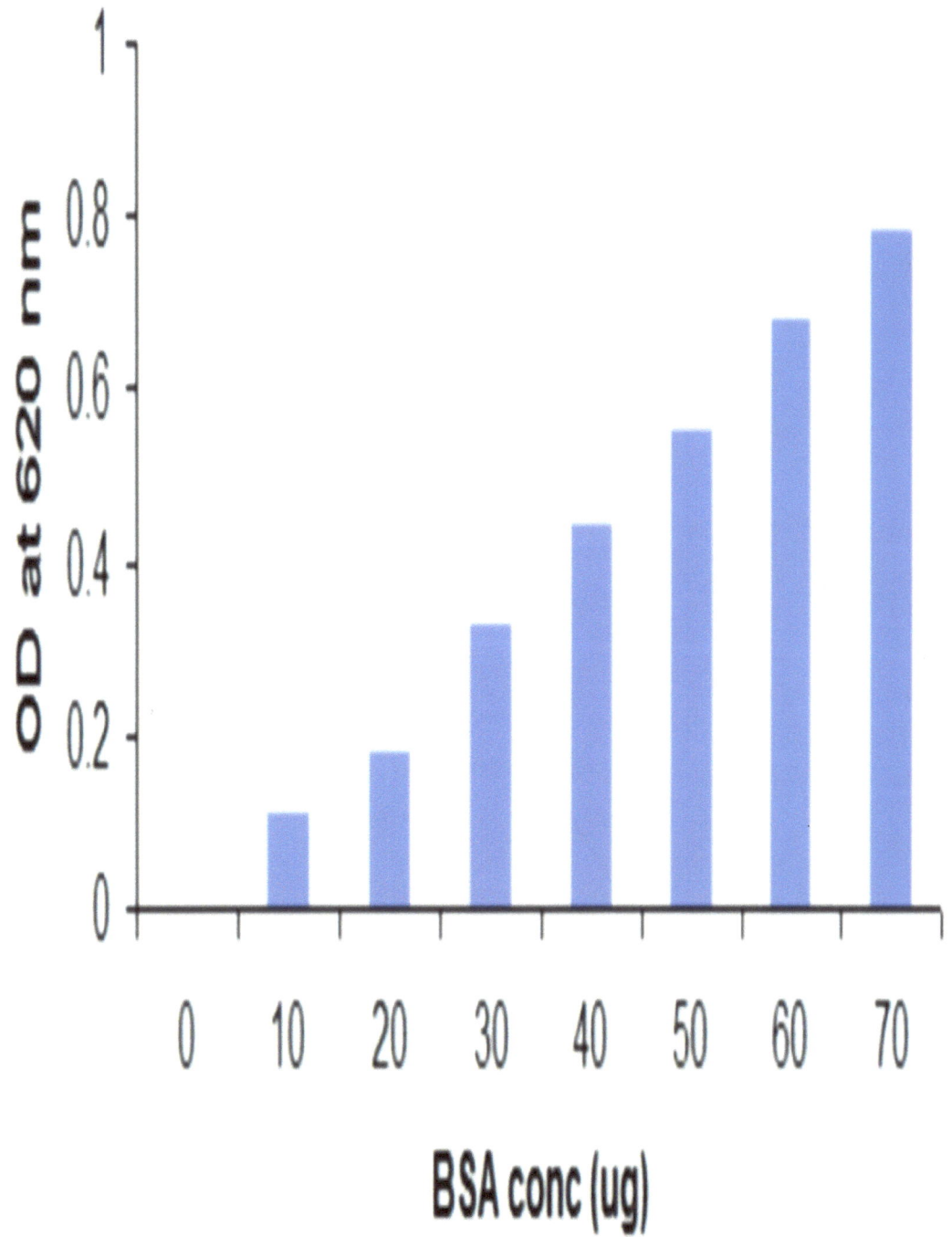

Fig 26. Estimation of Protein Taking BSA as Standard by Lowery et al Method

EXPPERIMENTAL:

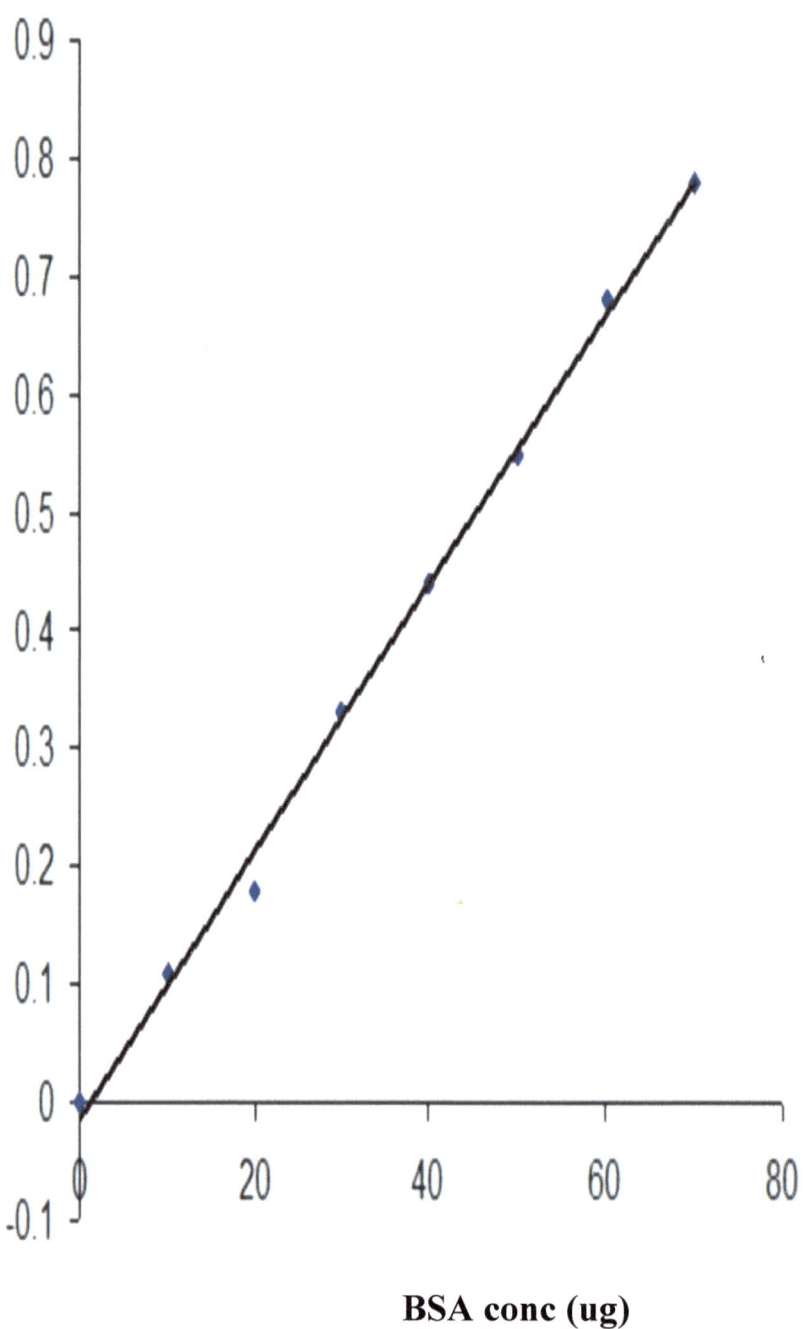

BSA conc (ug)

Fig 27. Estimation OF Protein OF Latex (Biological sample) by *Lower et. al* Method

STANDARD:

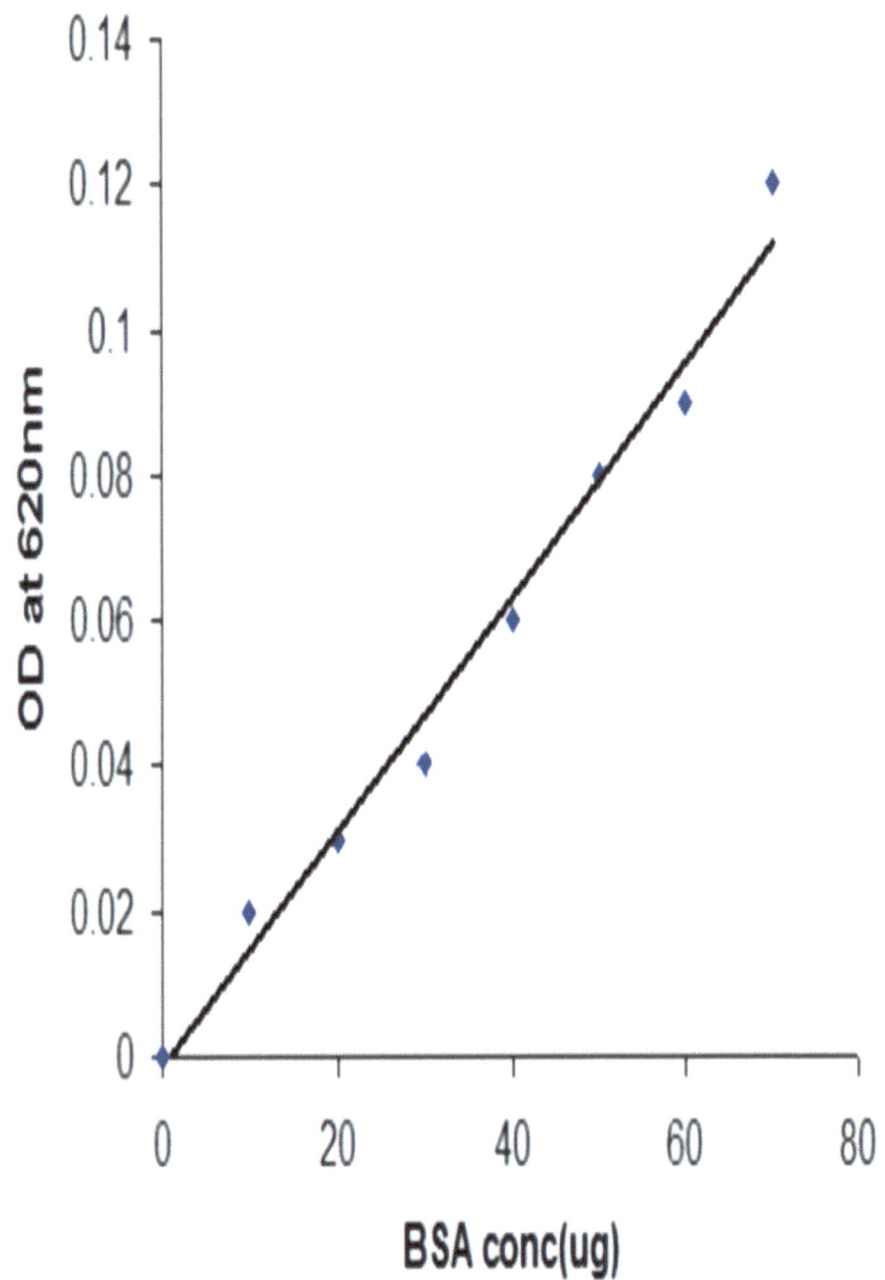

Fig28. Estimation OF Protein by *Lower et. al* Method

Experiment No.9

Objective: Estimation of Protein by **Biuret Method**

PRINCIPLE:

Biuret method was first described by A.G. Gornall. Biuret reagent consists of alkaline copper sulphate solution containing sodium potassium tartarate. The cupric ions form a coordination complex with four –NH group present in peptide bonds giving an absorption maximum at 540nm-560nm wavelength. Since the method is based on peptide bonds, it is absolute and very reproducible.

REACTION:

Cu^{2+}-sulfate in alkaline tartrate reacts with peptide bonds to produce a **Cu^{2+}-sulfate in alkaline tartrate** reacts with peptide bonds to produce a **purple compound** (copper-protein complex) with maximum absorption at 540 nm.

Reactions involved:

1. chelation

$$Cu^{2+}/OH^- + \text{protein} \longrightarrow [Cu^{2+}\text{-protein complex}]$$

2. redox reaction

$$Cu^{2+} + (\text{polar amino acids, Trp, Tyr})_{red} \longrightarrow Cu^+ + (\text{amino acids})_{ox}$$

PREPARATION OF REAGENTS

1. 10 % Sodium Hydroxide: For 150ml,
15 gm sodium hydroxide was dissolved in Glass Distilled water and volume was made to 150 ml with ditilled water.

2. 0.1N Sodium Hydroxide (500ml):
2 gm of sodium hydroxide was dissolved in Glass Distilled water and made upto 500 ml.

3. Biuret Reagent (500ml): 75 gm Copper sulphate and 3 gm Rochelle salt dissolved Glass distilled water and volume was made to 250 ml. To it was added 150 ml of 10 % NaOH solution. Raised the volume upto 500ml with Glass double distilled water.

4. Standard Protein Solution (10mg/ml); 100 ml : 1gm of egg albumin was dissolved in 20 ml 0.1N NaOH with very gentle stirring. Then volume rose to 100 ml with 0.1 N NaOH.

Experimental Procedure :

Seven test tube taken and arrange in a series from 1to seven. Aliquot of 0.1ml protein solution were taken to get a corresponding range upto 0.6 ml of protein. The volume in each test tube was raised upto 1ml by 0.1 N NaOH. Then 4 ml of biuret reagent was added in each of the test tube. Then test tube were incubated at 37^0 C for 30 minutes. After this reading is taken 560nm.

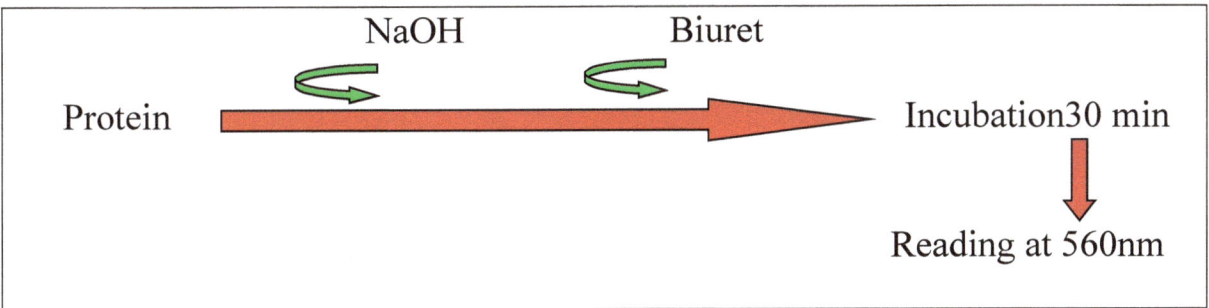

Fig 29: Protein Estimation By Biuret Method

Concentration of egg Albumin	Optical Density
0	0
10	0.032
20	0.064
30	0.1
40	0.12
50	0.16
60	0.19
70	0.23

Table4: Optical density at different concentration

Fig30. Estimation of protein by Biuret Method at different concentration.

Experiment No 10

Experiment: The isolation of casein from milk

Principle:
Casein is the main protein found in milk and is present at a concentration of about 35g/ litre. It is actually a heterogeneous mixture of phosphorus containing proteins and not a single compound.

Most protein show a minimum solubility at their isoelectric point and this principle is used to isolate the casein by adjusting at their isoelectric point and this principle is used to isolate the casein by adjusting the pH of the milk to 4.8 its isoelectric point. Casein is insoluble in ethanol and this property is used to remove unwanted fat from the preparation

Materials
- Milk
- Sodium acetate buffer (0.2 mol / litre, pH 4.6)
- Ethanol(95 % v /v)
- Ether
- Thermometer 100 °C
- Muslin
- Filter equipment and papers

Method:

Place 100 ml of milk in a 500 ml beaker and warm to 40 °C, also warm 100 ml of the acetate buffer and slowly with stirring The final pH of the mixture should about 4.8 and this can be checked with a pH meter. Cool the suspension to room temperature then leave to stand for further 5 min before filtering through muslin.

Wash the precipitate several times with a mixture with small volume of water then suspend it in about 30 ml of ethanol. Filtre the suspension on a funnel and wash the precipitate with equal amount of ethanol and ether. Finally wash the precipitate on the filter paper with 50 ml of ether and suck dry. Remove the powder and spread out on a watch glass to allow evaporation of the ether.

Weigh the casein and calculate the percentage yield of the protein.

Experiment No. 11

Objective: Estimation of Protein by Bradford protein assay

Introduction:

The Bradford protein assay is done to determine protein concentrations in solutions that depends upon the change in absorbance in Coomassie Blue G-250 upon binding of protein (Bradford, Anal. Biochem. 72: 248, 1976). Unlike many other assays, including the Lowry procedure, the Bradford assay is not susceptible to interference by a wide variety of chemicals present in samples. The assay is designed for use in microtiter plates.

Materials:

Bradford Reagent - The reagent can be made by dissolving 100 mg Coomassie Blue G-250 (available from several sources) in 50 ml 95% ethanol, adding 100 ml 85% (w/v) phosphoric acid to this solution and diluting the mixture to 1 liter with water.

Bovine serum albumins (BSA) (1 mg/ml) - BSA dissolve in saline and store it frozen in 1 ml aliquots for quick use. The standard should be dissolved in a buffer similar to that the unknowns will be dissolved in.

Microtiter plates -

Multi-well microtiter plate reader equipped with a 595 nm filter.

Procedure

Assay performed in duplicate or triplicate

1. Pipette 0, 2, 4, 6, 10, 15 and 20 µl of BSA (1 mg/ml) into assigned wells of a 96-well plate.
2. Pipette up to 20 µl of unknown samples into individual wells of a 96-well plate.
3. Add 40 µl of Bradford Reagent into all wells containing standard or sample.
4. Add dd H_2O to all wells to bring the final volume to 200 µl.
5. Read absorbance at 560 nm without any prior incubation.

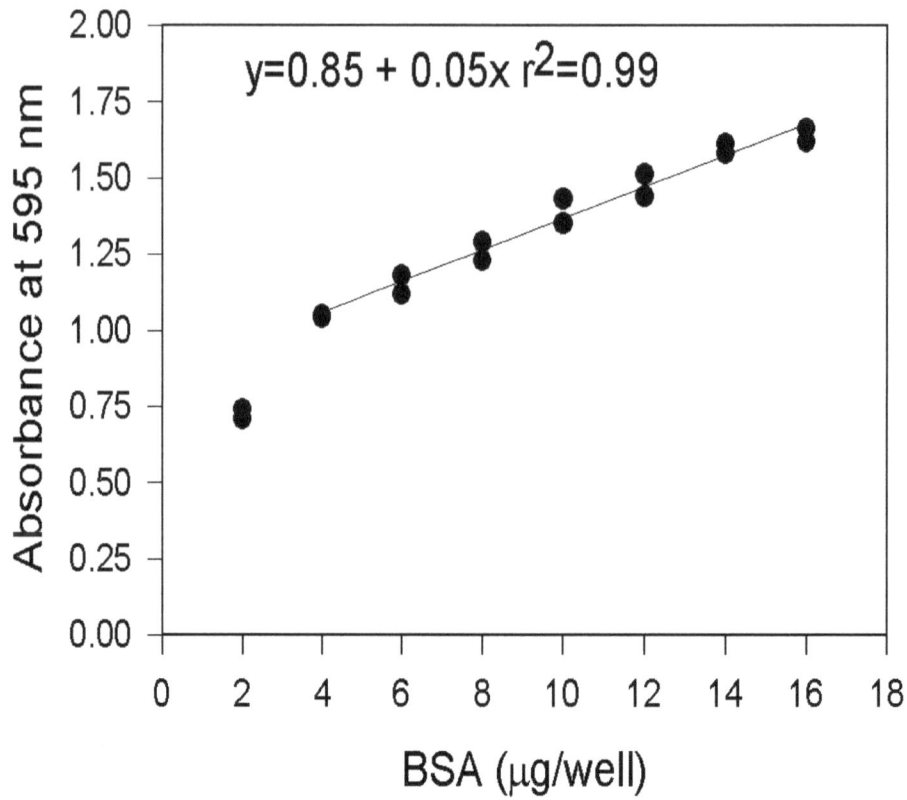

Fig 31. Estimation of Protein by Bradford protein assay

Fig 32. Bradford Reagent

Experiment No 12.

Objective: Estimation of Protein By KJELDAHL method

Principle:

KJELDAHL method based on sulfuric acid mineralization and measurement of the released ammonia. This method was invented by a Danish chemist, Johan G.C.T. Kjeldahl in 1883. Analyzing the content of nitrogenous compounds in urine, serum, or other biological specimens, usually to determine relatively large amounts of nitrogen (e.g., 20 to 100 mg).

Procedure: Treated with a digestion mixture (copper sulfate and sulfuric acid), heated and made alkaline with a solution of sodium hydroxide. Distilled ammonia from the mixture, trapped it in a boric acid-indicator solution and titrated with standard hydrochloric or sulfuric acid.

Kjeldahl method: The Kjeldahl method was developed in 1883. In this method food is digested with a strong acid so that it releases nitrogen which can be determined by a suitable titration technique. The amount of protein present is then calculated from the nitrogen concentration of the food. The same basic approach is still used today, although a number of improvements have been made to speed up the process and to obtain more accurate measurements. It is usually considered to be the standard method of determining protein concentration.

Because the Kjeldahl method does not measure the protein content directly a conversion factor (F) is needed to convert the measured nitrogen concentration to a protein concentration. A conversion factor of 6.25 (equivalent to 0.16 g nitrogen per gram of protein) is used for many applications, however, this is only an average value, and each protein has a different conversion factor depending on its amino-acid composition.

The Kjeldahl method can be divided into three steps:

1. Digestion,
2. Neutralization
3. Titration.

Principles:

1. Digestion:

The food sample to be analyzed is weighed into a digestion flask and then digested by heating it in the presence of sulfuric acid (an oxidizing agent which digests the food), anhydrous sodium sulfate (to speed up the reaction by raising the boiling point) and a catalyst, such as copper, selenium, titanium, or mercury (to speed up the reaction). Digestion converts nitrogen in the food (other than that which is in the form of nitrates or nitrites) into ammonia, and other organic matter to CO_2 and H_2O. Ammonia gas is not liberated in an acid solution because the ammonia is in the form of the ammonium ion (NH_4+) which binds to the sulfate ion (SO_4^{2-}) and thus remains in solution:

$$N \text{ (food)} \text{---------} (NH_4)_2SO_4 \quad (1)$$

2. Neutralization:

After the digestion has been completed the digestion flask is connected to a recieving flask by a tube. The solution in the digestion flask is then made alkaline by addition of sodium hydroxide, which converts the ammonium sulfate into ammonia gas:

$$(NH_4)_2SO_4 + 2\,NaOH \Longrightarrow 2NH_3 + 2H_2O + Na_2SO_4 \quad (2)$$

The ammonia gas that is formed is liberated from the solution and moves out of the digestion flask and into the receiving flask - which contains an excess of boric acid. The low pH of the solution in the receiving flask converts the ammonia gas into the ammonium ion, and simultaneously converts the boric acid to the borate ion:

$$NH_3 + H_3BO_3 \text{(boric acid)} \Longrightarrow NH_4+ + H_2BO_3- \text{ (borate ion)} \quad (3)$$

3. Titration:

The nitrogen content is then estimated by titration of the ammonium borate formed with standard sulfuric or hydrochloric acid, using a suitable indicator to determine the end-point of the reaction.

$$H_2BO_3^- + H^+ \text{----------} H_3BO_3 \quad (4)$$

The concentration of hydrogen ions (in moles) required to reach the end-point is equivalent to the concentration of nitrogen that was in the original food (Equation 3). The following equation can be used to determine the nitrogen concentration of a sample that weighs m grams using a xM HCl acid solution for the titration:

$$\% N = \frac{x \text{ moles}}{1000 \text{ cm}^3} \times \frac{(v_s - v_b) \text{ cm}^3}{m \text{ g}} \times \frac{14 \text{ g}}{\text{moles}} \times 100$$

Where vs and vb are the titration volumes of the sample and blank, and 14g is the molecular weight of nitrogen N. A blank sample is usually ran at the same time as the material being analyzed to take into account any residual nitrogen which may be in the reagents used to carry out the analysis. Once the nitrogen content has been determined it is converted to a protein content using the appropriate conversion factor: % Protein = F(conversion factor) %N.

Advantages:

- The Kjeldahl method is widely used internationally and is still the standard method for comparison against all other methods.

- Its universality, high precision and good reproducibility have made it the major method for the estimation of protein in foods.

Disadvantages:

- It does not give a measure of the true protein, since all nitrogen in foods is not in the form of protein. Different proteins need different correction factors because they have different amino acid sequences.

- The use of concentrated sulfuric acid at high temperatures poses a considerable hazard, as does the use of some of the possible catalysts

- The technique is time taking

Experiment No 13

Enhanced Dumas method For Protein Estimation

Objective: Estimation of Protein By Enhanced Dumas method

Introduction:

This is an automated instrumental technique capable of rapidly measuring the protein concentration of food samples. This technique is based on a method described by a scientist called **Dumas**. It is comparatively faster than Kjeldahl method and is also the standard method for analysis for proteins for some foodstuffs due to its rapidness.

Principles:

A sample of known mass is combusted in a high temperature (about 900 °C) chamber in the presence of oxygen. This leads to the release of CO_2, H_2O and N_2. The CO_2 and H_2O are removed by passing the gasses over special columns that absorb them. The nitrogen content is then measured by passing the remaining gasses through a column that has a thermal conductivity detector at the end. The column helps separate the nitrogen from any residual CO_2 and H_2O that may have remained in the gas stream. The instrument is calibrated by analyzing a material that is pure and has a known nitrogen concentration, such as EDTA (= 9.59 %N). Thus the signal from the thermal conductivity detector can be converted into a nitrogen content. As with the Kjeldahl method it is necessary to convert the concentration of nitrogen in a sample to the protein content, using suitable conversion factors which depend on the precise amino acid sequence of the protein.

Advantages:

- It is much faster than the Kjeldahl method (under 4 minutes per measurement, compared to 1-2 hours for Kjeldahl).
- It doesn't need toxic chemicals or catalysts.
- Many samples can be measured automatically.
- It is easy to use.

Disadvantages:

- High initial cost.

- It does not give a measure of the true protein, since all nitrogen in foods is not in the form of protein.

- Different proteins need different correction factors because they have different amino acid sequences.
- The small sample size makes it difficult to obtain a representative sample.

Protein Separation

Proteins can be separated by exploiting differences in their solubility in aqueous solutions. The solubility of a protein molecule is determined by its amino acid sequence. Proteins can be separated on basis of size, shape, hydrophobicity and electrical charge. Proteins can be selectively precipitated by changing the pH, ionic strength, dielectric constant or temperature of a solution. These separation techniques are easy and convenient to use when large quantities of sample are involved, because they are relatively quick, inexpensive.

Salting out:

Proteins can be precipitated from aqueous solutions when the salt concentration exceeds a critical level, which is known as salting-out, because all the water is "bound" to the salts, and is therefore not available to hydrate the proteins. Ammonium sulfate $[(NH_4)_2SO_4]$ is generally used because it has a high water-solubility, other neutral salts may which can be are sodium chlorids or potassium chloride(NaCl or KCl). This is two-step procedure for separation efficiency.

In the first step, the salt is added at a concentration just below that necessary to precipitate out the protein of interest. The solution is then centrifuged to remove any proteins that are less soluble than the protein of interest.

The salt concentration is then increased to a point just above that required to cause precipitation of the protein. This precipitates out the protein of interest (which can be separated by centrifugation), but leaves more soluble proteins in solution. The main problem with this method is that large concentrations of salt contaminate the solution.

Isoelectric Precipitation:

The isoelectric point (pI) of a protein is the pH where the net charge on the protein is zero. Proteins tend to precipitate at their pI because there is no electrostatic repulsion keeping them apart. Proteins have different isoelectric points because of their different amino acid sequences (i.e., relative numbers of anionic and cationic groups), and thus they can be separated by adjusting the pH of a solution. When the pH is adjusted to the pI of a particular protein it precipitates leaving the other proteins in solution.

Solvent Fractionation:

The solubility of a protein depends on the dielectric constant of the solution because this alters the magnitude of the electrostatic interactions between charged groups. As the dielectric constant of a solution decreases the electrostatic interactions between charged species increases. This in leads to decrease in the solubility of proteins in solution because they are less ionized, and therefore the electrostatic repulsion between them is not sufficient to prevent them from aggregating. The dielectric constant of aqueous solutions can be lowered by adding water-soluble organic solvents, such as ethanol or acetone. The amount of organic solvent required to cause precipitation depends on the protein and therefore proteins can be separated on this basis. Solvent fractionation is usually performed at 0°C or below to prevent protein denaturation caused by temperature increases that occur when organic solvents are mixed with water.

Separation of Protein on the Basis of Different Adsorption Characteristics:

The separation of compounds by selective adsorption-desorption at a solid matrix that is contained within a column through which the mixture passes is the basis of
Adsorption chromatography. Separation is based on the different affinities of different proteins for the solid matrix

Adsorption chromatography is of two types:

Affinity Chromatography:

Affinity chromatography uses a stationary phase that consists of a ligand covalently bound to a solid support. The ligand is a molecule which has a highly specific and unique reversible affinity for a particular protein. The sample is passed through the column and the protein of interest binds to the ligand, whereas the contaminating proteins pass directly through the column. The protein of interest is then eluted using a buffer solution which favors its desorption from the column. This technique is the most efficient means of separating a protein from a mixture of proteins, but it is most expensive technique, because of specific ligands bound to them.

Ion Exchange Chromatography: Ion exchange chromatography relies on the reversible adsorption-desorption of ions in solution to a charged solid matrix. This technique is the most commonly used for protein separation. A positively charged matrix is called an anion-exchanger because it binds negatively charged ions (anions). A negatively charged matrix is called a cation-exchanger because it binds positively charged ions (cations).

The buffer conditions (pH and ionic strength) are adjusted to favor maximum binding of the protein to the ion-exchange column. Contaminating proteins bind less strongly and are eluted through the column. The protein of interest is then eluted using another buffer solution which favors its desorption from the column (e.g., different pH or ionic strength).

Separation Due to Size Differences of Protein:

The size of protein is the basis of this kind of separation. The molecular weights of proteins vary from about 10,000 to 1,000,000 daltons. More precisely along with molecular weight the radius of protein molecules is the basis of separation. Proteins with the same molecular weight the radius increases in the following order: compact globular protein < flexible random-coil < rod-like protein.

Dialysis:

Dialysis is used to separate protein molecules in solution by use of semipermeable membranes. The membrane permit the passage of molecules smaller than a certain size through, but prevent the passing of larger molecules. A protein solution is placed in dialysis tubing which is sealed and placed into a large volume of water or buffer which is slowly stirred. Low molecular weight solutes flow through the bag, but the large molecular weight protein molecules remain in the bag. Dialysis is a relatively slow method.

Ultrafiltration:

In this method of separation pressure is applied. This method is similar to dialysis. A solution of protein is placed in a cell containing a semipermeable membrane, and pressure is applied.

Smaller molecules pass quickly through the membrane, whereas the larger molecules remain in the solution. That portion of the solution which is retained by the cell (large molecules) is called the retentate, whilst that part which passes through the membrane (small molecules) forms part of the ultrafiltrate.

Size Exclusion Chromatography:

This technique, sometimes known as gel filtration, also separates proteins according to their size. A protein solution is poured into a column which is packed with porous beads made of a cross-linked polymeric material (such as dextran or agarose).

Molecules larger than the pores in the beads are excluded, and move quickly through the column, whereas the movement of molecules which enter the pores is retarded. Thus molecules are eluted off the column in order of decreasing size. Beads of different average pore size are available for separating proteins of different molecular weights.

Molecular weights of unknown proteins can be determined by comparing their elution volumes V_o, with those determined using proteins of known molecular weight: a plot of elution volume versus log(molecular weight) should give a straight line.

Beer-Lambert Law

The Beer-Lambert Law states that the amount of light absorbed is proportional to the number of molecules of absorbing substance in the light path, ie. absorption is proportional both to the concentration of the sample solution and to the length of the light path through the solution. This relationship can be expressed as follows:

Absorbance, $A = \varepsilon \times c \times l$

c = concentration of the sample (in Moles/liter),
l = length of the light path through the solution (in cm) and
ε = molar extinction coefficient

To determine the absolute concentration of a pure substance, a standard curve is constructed from the known concentrations and using that standard curve, the absorbance reading of the unknown concentration was determined. The determination of unknown concentration from the standard curve is done by drawing a line parallel to the X- axis from the point on the Y axis that corresponds to the absorbance of the unknown.

This line will be made to intersect the standard curve drawn, and is extended vertically such that it meets the X-axis and the concentration of unknown is read from the X-axis. A typical standard curve is depicted in the figure.

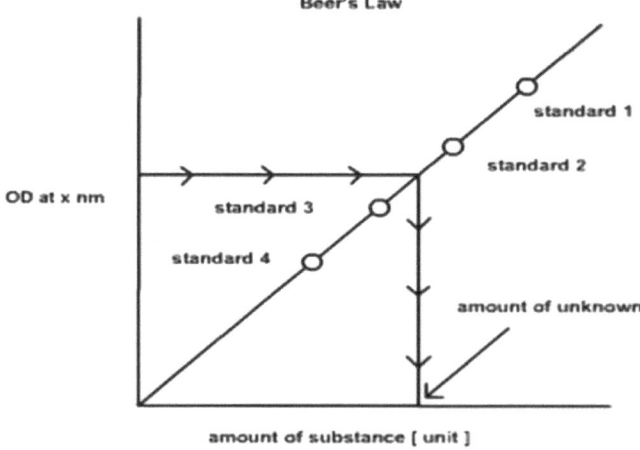

Fig 33. Graph depicting Beer-Lambert Law

Principle of the Colorimeter:

Unknown compounds may be identified by their characteristic absorption spectra in the ultraviolet, visible or infrared regions. Enzyme-catalysed reactions frequently can be followed by measuring spectrophotometrically the appearance of a product or disappearance of a substrate. A spectrophotometer /colorimeter is an instrument for measuring the **absorbance** of a solution by measuring the amount of light of a given wavelength that is transmitted by a sample.

Light can be categorized according to its wavelength. Light in the short wavelengths of 200 to 400 nm is referred to as ultraviolet (UV). Light in the longer wavelengths of 700 to 900 nm is referred to as near infrared (near IR).

Visible light falls between the wavelengths of 400 and 700 nm. All the colors visible to human eye falls under this wavelength range. Any solution that contains a compound that absorbs light in the visible region will appear colored to the eye. The solution is colored because specific wavelengths of light are absorbed as they pass through the solution. Then, the only light that the eye will perceive are the wavelengths of light that are transmitted (not absorbed)

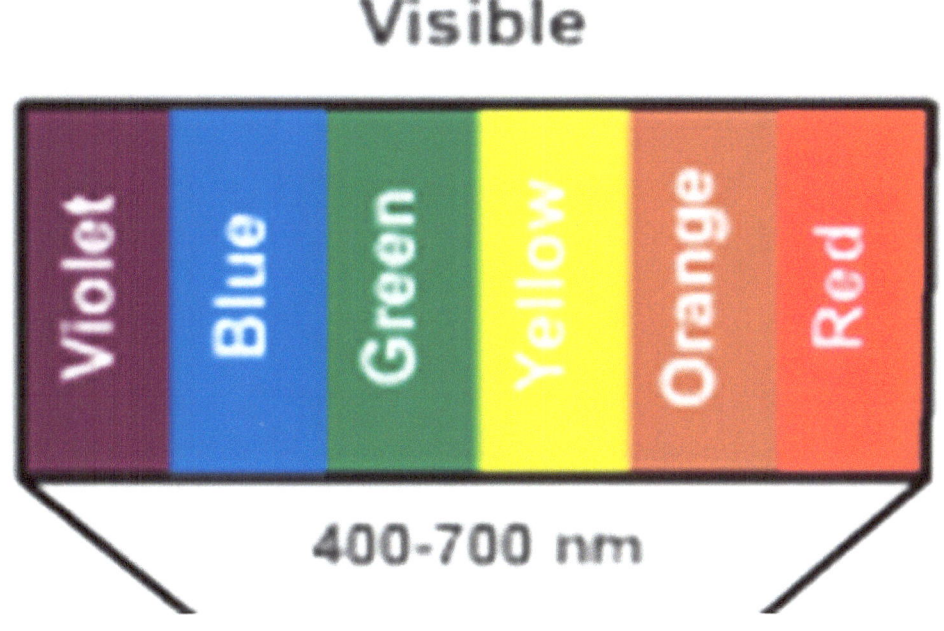

Fig34. Visible Light Spectrum from wavelength 400-700nm

All the colors visible to human eye come under this wavelength range. Solution that contains a compound that absorbs light in the visible region will appear colored to the eye. The solution appear red is colored because specific wavelengths of light are absorbed as they pass through the solution.

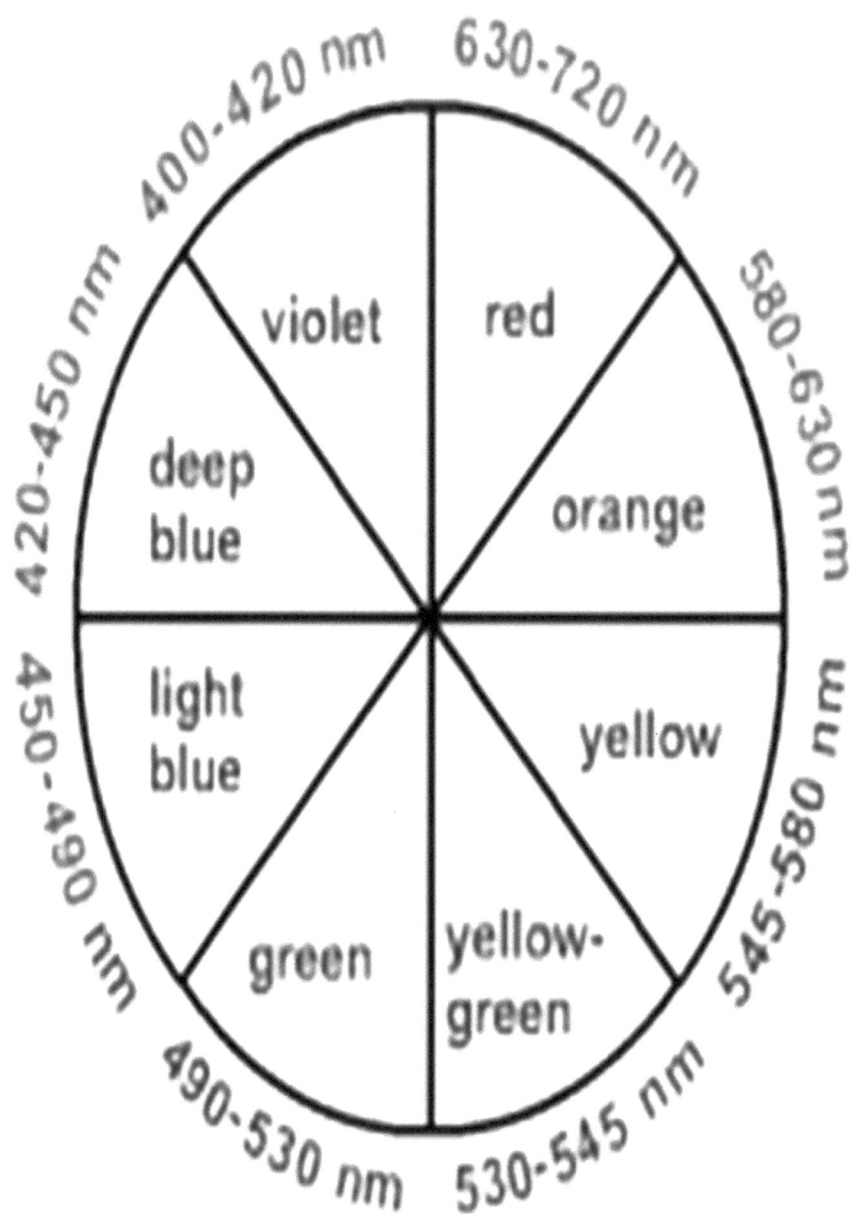

Fig35. Different color of light absorb different wavelength of light.

Working OF The Colorimeter

1. Switch on the colorimeter and set the wavelength at which the absorbance is to be measured [30 min before taking the readings].

2. Rinse the cuvette with the blank solution, drain off the solution by inverting the cuvette and touching its mouth on to the filter paper.

3. Fill the cuvette with the blank solution up to the appropriate mark and place it in the hole provided on the colorimeter.

4. Care should be taken that the cuvette is properly inserted into the hole.

5. Read the OD value and adjust the instrument to zero.

6. Take out the cuvette and drain off the blank solution.

7. Now measure the OD values from the lower to the higher concentration of the sample taken.

8. After taking reading cuvette is properly washed with distilled water dried and kept back in place.

9. Care should be taken to Swtich off the plug of colorimeter before leaving the lab.

Precaution taken During Working On Colorimeter:

- While operating the colorimeter, make sure that water droplets do not enter the slot of the colorimeter as it will damage the instrument.

- So wipe the cuvette using a lint free tissue paper before inserting it into the slot.

- Do not touch the part of the cuvette that gets inserted into the colorimetric slot with hands and ensure that the cuvette is free from water droplets.

Comparative Analysis OF Three Method Of Protein Estimation

1. Modified Lowry:

- Range: 2 to 100 micrograms
- Volume: 1 ml (scale up for larger cuvettes)
- Accuracy: Good
- Major interfering agents: Strong acids, ammonium sulfate

2. Modified Biuret:

- Range: 1 to 10 mg
- Volume: 5 ml (scale down for smaller cuvettes)
- Accuracy: Good
- Major interfering agents: Ammonium salts

3. Bradford assay

- Range: 1 to 20 micrograms (micro assay); 20 to 200 micrograms (macro assay)
- Volume: 1 ml (micro); 5.5 ml (macro)
- Accuracy: Good
- Major interfering agents: None

Reagents Required

1. Ninhydrin reagent: Ninhydrin (2%W/V) in acetone

2. Sodium hydroxide (40%W/V)

3. Conc. Nitric acid

4. Hydrochloric acid (6N)

5. Sulphanilic acid (1%W/V) in 1N HCl

6. Sodium nitrite (5%W/V) in distilled water (to be freshly prepared)

7. Sodium carbonate (10%W/V) in distilled water

8. Millon's reagent (15%W/V mercuric sulphate in 6N sulphuric acid)

9. Sodium nitrite solution(5%) [To be freshly prepared]

10. Bromine (5%) in acetic acid(33%) solution

11. Amonium carbonate(5%)

12. Acetic acid – Glyoxilic acid reagent – Glacial acetic acid is exposed to sun light (for 5 – 6 hours) for the formation of small amounts of glyoxilic acid)

13. Con. Sulphuric acid

14. Sodium hydroxide (10% W/V)

15. α Naphthol reagent (1%W/V in ethyl alcohol)

16. Hypobromite solution (To be freshly prepared) : -Take 100 of 5%(W/V) sodium hydroxide solution in a glass reagent bottle and add 1ml of pre chilled liquid bromine, using a pro pipette. Shake the contents till bromine dissolves)

17. Urea solution 5% (W/V)

18. Lead acetate solution (10% W/V)

19. 5N NaOH

20. 1% Glycine solution

21. 10% Sodium nitroprusside solution

22. Isatin reagent : Isatin (1% W/V) in acetic acid

Lowry method

23. 0.1 N NaOH: Dissolve 4 g NaOH in glass distilled water and made the volume upto 1 liter with GD water.

24. Alkaline Copper Reagent:

(a) Sodium Carbonate solution (Na_2CO_3) (2%, w/v):   100ml.
Sodium carbonate 2g was dissolved in sodium hydroxide 0.1N and the volume made up to 100 ml

(b) Copper sulphate solution (1%, w/v):   100ml
Copper sulphate (1.0 g) was dissolved in glass distilled water and volume made to 100ml.

(c) Sodium Potassium Tartrate (Rochelle salt)(2%,w/v)   100ml

Sodium potassium tartrate 2g was dissolved in glass distilled water and the volume raised to 100 ml. Alkaline copper sulphate solution was prepared by mixing 100ml of solution (a) with 1.0 ml each of (b) and (c). The solution was prepared fresh daily just before use.

25. BSA (Bovine serum albumin)

26. Stock solution (200ug/ml):

Dissolved 20 mg of BSA in 0.1N NaOH and made the volume upto 100 ml with 0.1N NaOH. 0.1 ml this contain thus 200 µg of BSA. We got 1 mg /ml solution of BSA.

27. Working solution:

Took 10 ml of this stock solution and diluted to 50 ml 0.1 N NaOH sodium hydroxide solution.

28. Folin's Reagent :

Biuret Method

29. 10%Sodium Hydroxide: For 150ml,
    15 gm sodium hydroxide was dissolved in Glass Distilled water and volume was made to 150 ml with ditilled water.

30. 0.1N Sodium Hydroxide (500ml):
    2 gm of sodium hydroxide was dissolved in Glass Distilled water and made upto 500 ml.

31. Biuret Reagent (500ml):75 gm Copper sulphate and 3 gm Rochelle salt dissolved   Glass distilled water and volume was made to 250 ml. To it was added 150 ml of   10% NaOH solution.   Raised the volume upto 500ml with Glass double distilled water.

32. Standard Protein Solution (10mg/ml); 100ml : 1gm of egg albumin was dissolved   in 20 ml 0.1N NaOH with very gentle stirring. Then volume rose to 100 ml with 0.1   N NaOH.

    Casein

33. Milk
32. Sodium acetate buffer (0.2 mol / litre, pH 4.6)
33. Ethanol(95 % v /v)
34. Ether
35. Thermometer 100 °C
36. Muslin
37. Filter equipment and papers

Bradford

38. Bradford Reagent –
39. Bovine serum albumin (BSA) (1 mg/ml) –

List of Materials Required

1. Test tube
2. Test tube holder
3. Test tube stand
4. Spatula
5. Water bath
6. Dropper
7. Pipette
8. Vortex mixer
9. Filter paper strip
10. Ice box
11. Digital weighing machine
12. Glass rod
13. Beaker
14. Wash Bottle
15. Colorimeter
16. Volumetric Flask
17. Flat Bottom Flask
18. Boiling test tube
19. Watch glass

Fig 36. Beaker

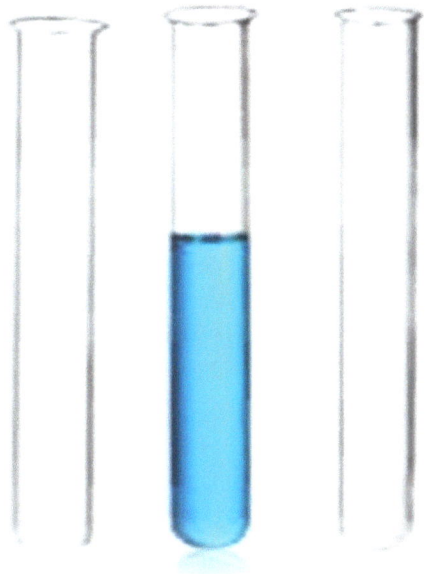

Fig 37. Test Tube

Fig 38. Flat Bottom Conical Flask

Fig 39. Test Tube Stand

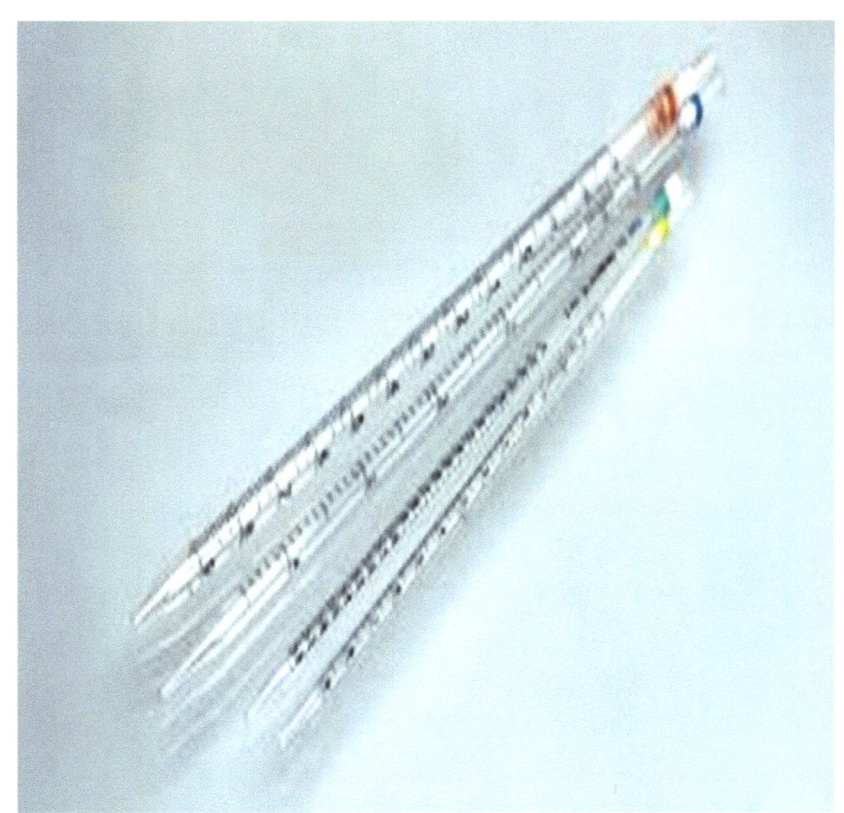

Fig 40. Peppette

Fig 41. Test Holder

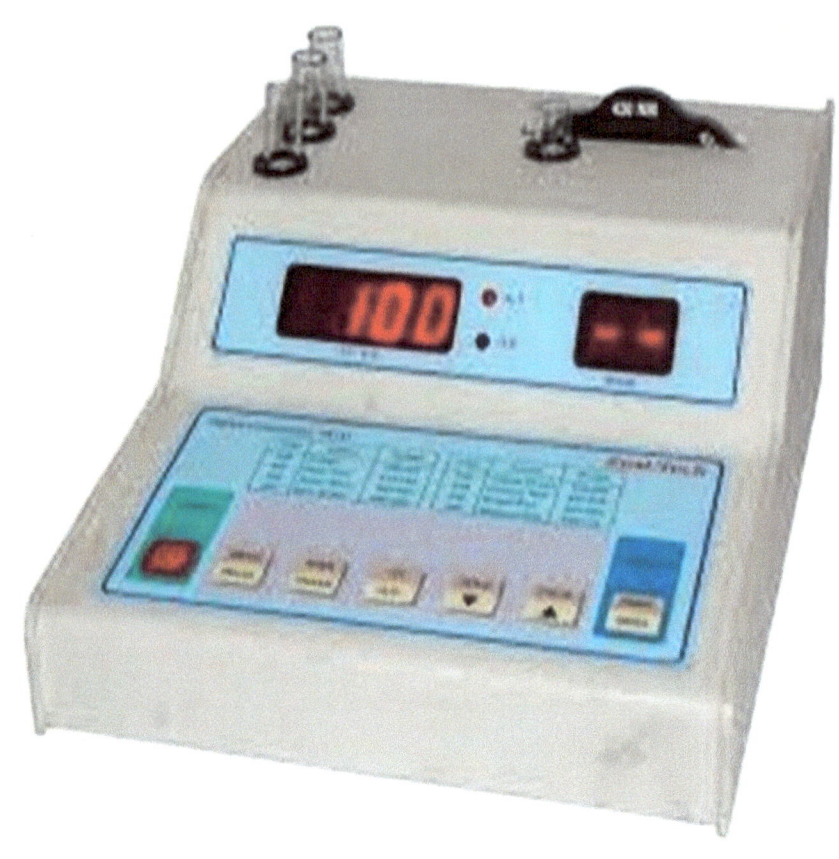

Fig 42. Colorimeter

Fig 43. Vortex

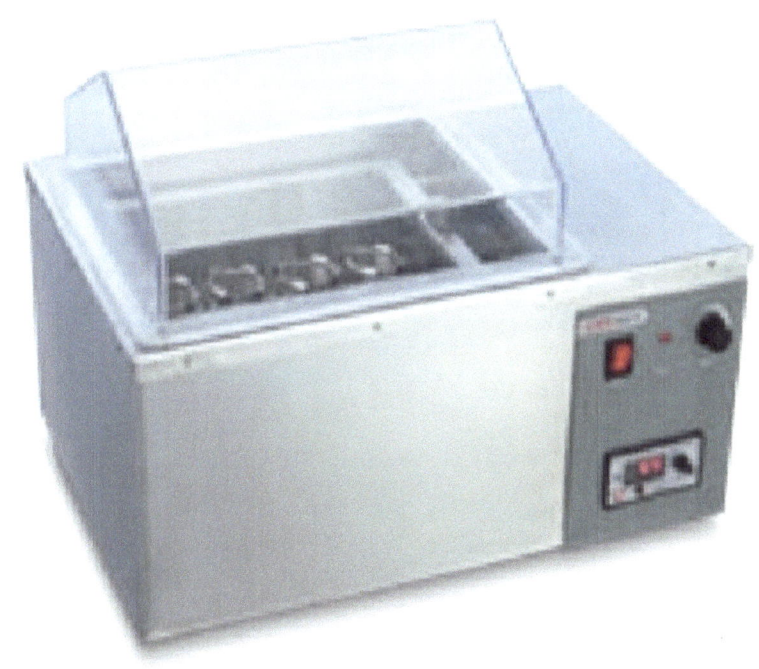

Fig44. Waterbath

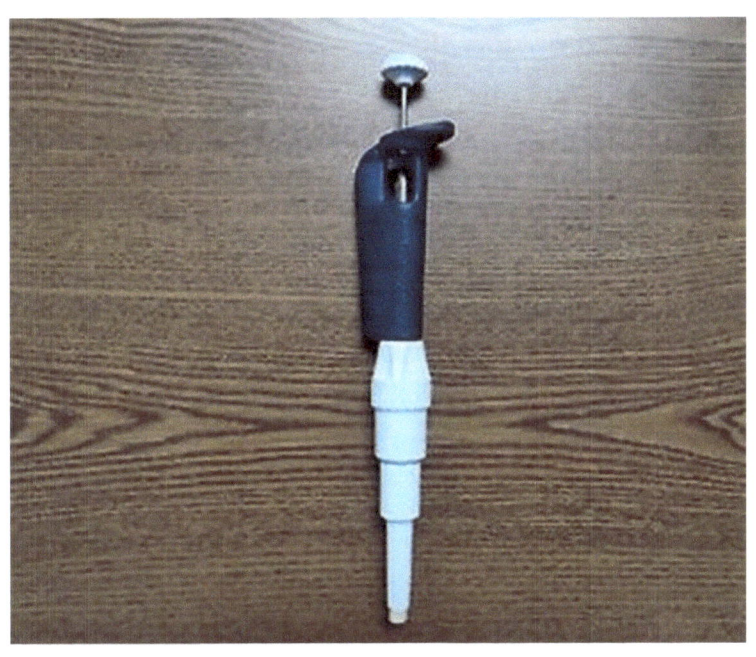

Fig 45. Pipette

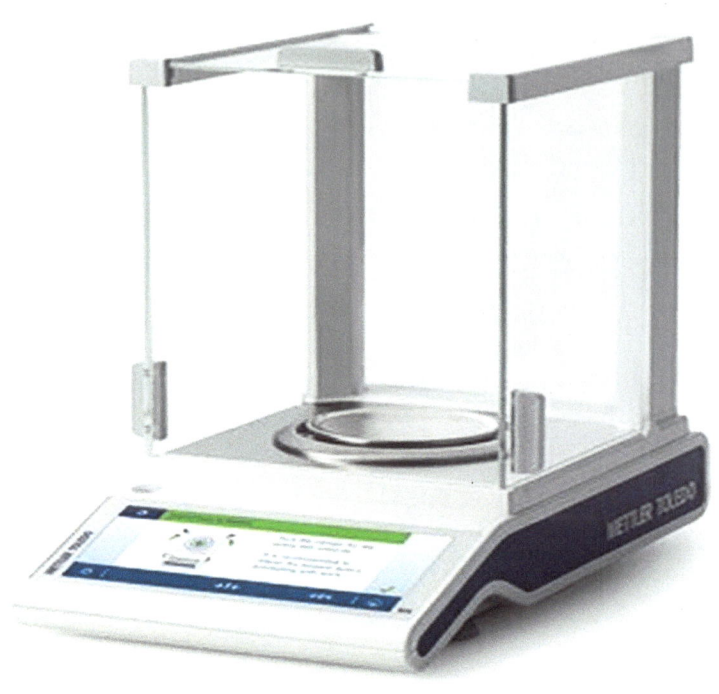

Fig 46: Analytical Balance

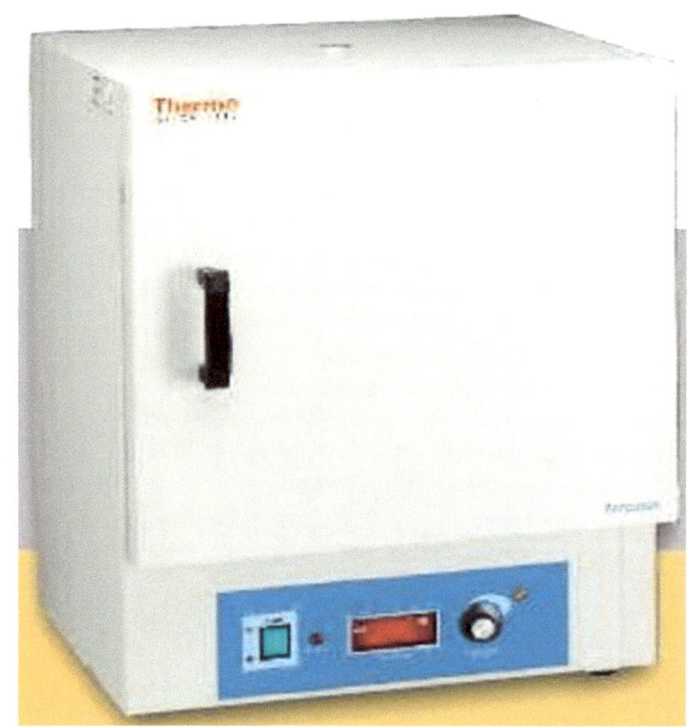

Fig47. Oven

PRECAUTIONS IN LAB:

1. Always wear lab coat and gloves when you are in the lab.
2. Switch on the exhaust fan when you work in lab.
3. Care should be taken while handling reagents like Conc.Sulphuric acid and Hydrochloric acid. These concentrated acids should be opened and used only in a FUMEHOOD. These concentrated acids cause severe burns and on inhaling can cause damage to the lining of the lungs.
4. Reagents like Ninhydrin reagent, sulphanilic acid, isatin reagent, bromin, Sodium nitroprusside should also be handled with care. Accidental spill of these reagent will cause burns and itches. Wash the spilled area with cold water and inform the lab assistant immediately.
5. Make sure that the water bath is set to the proper temperature before starting with the experiment.
6. Take care while heating the sample over the flame.
7. In Xanthoproteic test, results can be observed clearly on boiling the contents in a waterbath.
8. Wipe the lab bench after the experiment is completed.
9. Make sure to switch off the waterbath and the exhaust fans before leaving the lab.

www.ingramcontent.com/pod-product-compliance
Lightning Source LLC
Chambersburg PA
CBHW050749180526
45159CB00003B/1398